澜湄职业教育培训中心暨柬埔寨鲁班工坊系列教材

A Series of Textbooks for Lancang-Mekong Vocational Education Training Center and Cambodia Luban Workshop

4G 通信网络管理员(初级)

4G Communication Network Administrator(Primary)

主 编 高 源

Chief editor GAO Yuan

副主编 刘赟宇 韩 健 胡 炜

Deputy editors LIU Yunyu HAN Jian HU Wei

西安电子科技大学出版社

Introduction to the Course

Based on TD–LTE technology, this book adopts a learning mode that features modularized training, and chooses the eNodeB made by Datang Telecom Group as learning platform. Hardware structures, installation of the host, and antenna connection of 4G base station are explained in a comprehensive and detailed way through specific task modules. The book is divided into 3 training modules including Introduction to TD–LTE Base Station and EPC Network, Installation of Base Station Equipment, and Installation of Auxiliary Equipment in Base Station and Optical Fiber Fusion Splicing. Readers can carry out correct installation and connection of 4G base station equipment by completing these three training modules. The purpose of this book lies in cultivating high-quality and skilled personnel who can construct and maintain 4G base station, helping them master the skills required for operating and maintaining communication base station, and meeting the requirements for standardized on-site maintenance.

Rich in content, the book explains in detail the installation process of 4G base station equipment in a standard way, and illustrates the process by using the on-site construction data provided by Datang. The book is highly practical, for it focuses on the work processes and the actual tasks.

The book can be used as the textbook of 4G communication course for electronics, communication and related majors in higher vocational colleges, as well as a reference book for professionals and engineering technicians who are interested in 4G communication technology.

图书在版编目(CIP)数据

4G 通信网络管理员：初级 = 4G Communication Network Administrator(Primary) / 高源主编. —西安：西安电子科技大学出版社，2021.12
ISBN 978-7-5606-6062-2

Ⅰ. ① 4… Ⅱ. ①高… Ⅲ. ①第四代移动通信—计算机网络管理—资格考试—自学参考资料 Ⅳ. ①TN929.537

中国版本图书馆 CIP 数据核字(2021)第 249132 号

策划编辑　刘玉芳
责任编辑　任倍萱　刘玉芳
出版发行　西安电子科技大学出版社(西安市太白南路 2 号)
电　　话　(029)88202421　88201467　　邮　编　710071
网　　址　www.xduph.com　　　　　电子邮箱　xdupfxb001@163.com
经　　销　新华书店
印刷单位　咸阳华盛印务有限责任公司
版　　次　2021 年 12 月第 1 版　　2021 年 12 月第 1 次印刷
开　　本　787 毫米 × 1092 毫米　1/16　　印　张　5.5
字　　数　117 千字
印　　数　1～1000 册
定　　价　21.00 元

ISBN 978-7-5606-6062-2 / TN
XDUP 6364001-1
***** 如有印装问题可调换 *****

General Foreword

Serving the Belt and Road Initiative of China, the Lancang-Mekong Vocational Education Training Center and Cambodia Luban Workshop is a joint project undertaken by Tianjin Sino-German University of Applied Sciences(TSGUAS) for the Ministry of Foreign Affairs, the Ministry of Education and the Tianjin Municipal People's Government. Based in Cambodia, the project is designed to serve five countries in the Lancang-Mekong area and radiate to other ten ASEAN countries. It integrates functions of vocational training, vocational education, scientific research, cultural inheritance and innovation&entrepreneurship, develops both academic and non-academic education, and operates as a market-oriented international vocational training center.

At the initial stage of the project, 18 training rooms including mechanical processing technology, electrical technology and communication technology were built in three training centers for mechatronics and communication technology majors, with a total construction area of 6,814 m^2 and more than 1,600 sets of equipment.

The project will implement a "three-phase" plan. Based on the specialty construction in the first phase, international tourism, logistics engineering, automobile maintenance, building electricity and other specialties will be set up in the second phase to carry out technical skills training for Chinese&Cambodian enterprises and Cambodian people. Meanwhile, higher vocational education, applied technology undergraduate education, joint postgraduate education and other academic educations will be carried out to explore systematic talents cultivation of "medium and high vocational education, undergraduate education, and postgraduate education for a master's and doctoral degree".

Since 2017, as many as 95 articles about the project have been published by mainstream media including People's Daily, Guangming Daily, China Education News, Xinhuanet, etc. from home and abroad. After over two months of field study and research, Tianjin Television produced two feature stories named "Khmer Training", each lasting 30 minutes. The two episodes were broadcast on May 6th and May 13th 2019 respectively, featuring "on and on sails the vocational education, overseas shines the Luban Workshop". They give a full coverage of how TSGUAS teachers brought advanced skills to local areas and how friendship flourished along the Belt and Road Initiative route—a great contribution to the BRI. On July 18, 2019, the Royal Government of Cambodia conferred the Officer of the SAHAMETREI Medal to the Secretary of the Party Committee of TSGUAS, and the Knight of the SAHAMETREI Medal to the President and Vice President in charge of this project, with the signature of Prime Minister

Hun Sen of Cambodia. On July 22, 2019, China Education Association for International Exchange awarded TSGUAS the medal of "Featured Cooperation Project of China-ASEAN Higher Vocational Colleges". In October 2019, the President of National Polytechnic Institute of Cambodia (NPIC) presented 11 teachers with certificates and medals for their outstanding contributions to the Ministry of Labor and Vocational Training of Cambodia. Tianjin Sino-German University of Applied Sciences together with National Polytechnic Institute of Cambodia (NPIC) and their partners with enterprises was approved as the Belt and Road Joint Laboratory (Research Center)—Tianjin Sino-German and Cambodia Intelligent Motion Device and Communication Technology Promotion Center in December, 2020.

The Center has become a training base in Langcang-Mekong areas for technical talents training, a talent support base for Chinese enterprises overseas, a demonstration base for international students, and a base for teachers training. The Center is a key educational project of the Ministry of Foreign Affairs to serve the Belt and Road Initiative with foreign participation and entity institutions involved locally. The project will serve the social-economic and cultural development of the countries along the Initiative, enhancing the well-being of mankind; it will also serve the production output capacity of Chinese enterprises to help national development as well as enhance the international development of vocational education and the quality of its connotation. The project is a bridge connecting vocational education of Tianjin with the world, which marks a new stage of the city's international exchange and cooperation from a lower-medium to a medium-higher level.

The team of the project has compiled a series of textbooks for training, involving six occupations (electrotechnics, lathe, milling, CNC operation, bench and 4G communication network) from primary, intermediate to senior based on current human resources situation in Langcang-Mekong countries, China's teaching equipment, and Chinese vocational qualification standards. These 19 textbooks target competence development and orient students to work tasks, combining theory with practice, and learning with practicing so as to put knowledge and skills into real situations. The textbooks aim to provide skills standards for the six occupations and lay foundations for the upgrading of the technological level of Lancang-Mekong countries.

ZHANG Xinghui

Party Secretary of Tianjin Sino-German University of Applied Sciences

March, 2021

A Series of Textbooks for Lancang-Mekong Vocational Education Training Center and Cambodia Luban Workshop Editorial Committee

Preface

This book, based on the idea of resolving projects into tasks, is oriented toward practical operation, and the content is compiled according to tasks required by the post. It focuses on the network installation, network maintenance, network testing and network optimization of the 4G mobile communication system of the Datang Telecom Group, and is also integrated with skills training as well as competence training. It is creative in curriculum and selective in content.

In order to cultivate application-oriented talents in the field of mobile communication technology, the book is co-written by a group of elite lecturers with rich teaching experience, and senior engineers experienced with engineering practice through school-enterprise cooperation. According to the learning curve of students, and the position and level in the enterprise, this set of textbooks is divided into three volumes: 4G Communication Network Administrator – Primary (Level 5), 4G Communication Network Administrator – Intermediate (Level 4), and 4G Communication Network Administrator – Senior (Level 3).

In the Primary volume: Students will be trained to learn the basic engineering installation skills required for junior workers. The posts are installation engineers and installation supervisors.

In the Intermediate volume: After learning the Primary volume and having two years of work experience on the project site, students enter the Intermediate stage. In this period, students are going to operate, adjust and maintain the base station. The post is the test engineer, as well as the operation and maintenance engineer.

In the Senior volume: After learning the Intermediate volume and having one or two years of work experience on the project site, students enter the Senior stage. In this period, students are going to learn base station testing, network planning, and network optimization. The posts are planning engineers and optimization engineers.

This book is suitable to be used as the textbook for communication and electronic information related courses in application-oriented undergraduate and higher vocational education, and the training material for mobile communication and network technicians.

As a primary volume, this book, based on the mainstream product of Datang Telecom Group, the EMB5116 base station, combines the theory with the actual construction made by Chinese operators, and introduces installation and adjusting process of 4G LTE base station systematically and comprehensively, which provides a bridge between theory and practice for beginners with certain knowledge on communication, and readers interested in LTE network technology.

It is available to scan the two-dimensional code below for corresponding contents of this book in Chinese.

Due to the limited level of the author, there might be some unintentional mistakes in this book. Therefore, advice and correction from the readers are greatly appreciated.

Editor

Oct 2021

译文

Content

Training Module 1　Introduction to TD-LTE Base Station and EPC Network

[Brief description]

TD-LTE (Time Division-Long Term Evolution) products made by Datang group consist mainly of: base station (evolved Node Base, or eNB for short), Evolved Packet Core network (EPC for short), Operation and Maintenance Center (OMC for short). This project provides typical tasks designed for the personnel who are involved in the installation of mobile communication system. The tasks focus mainly on the introduction to and acquaintance with the equipment in those three parts as the training objectives. They cover the introduction to base station products, EPC network products, OMC network management products, as well as the installation of equipment. With the practical training in this project, the trainees are expected to meet the basic requirements for installing and adjusting the mobile network.

[Elements of the training]

1. Knowledge objectives

(1) Master the the technical characteristics, appearance, and software and hardware structure of mobile communication system products.

(2) Master the services and functions of mobile communication system products.

(3) Master typical configurations of mobile communication system products.

(4) Master the installation of mobile communication system products.

2. Ability objectives

(1) Master the basic composition of mobile communication system products.

(2) Master the services and functions of various mobile communication system products.

(3) Master the installation procedures of mobile communication system products.

3. Operational standards

This module focuses mainly on the introduction to the equipment, familiarizes students with the operation procedure and helps them master the operational standards.

4. Safety standards

In order to ensure safty for people and equipment during the experiment, people involved must strictly follow the operating rules listed below.

1) Illustration of the symbols

The types and meanings of different symbols are shown in Table 1-0-1.

Table 1-0-1　Types and Meanings of Different Symbols

Symbol	Meaning
⚠	Caution Symbol Indicating the general safety matters
(antistatic symbol)	Antistatic Symbol Indicating that the equipment is sensitive to static electricity
(high voltage symbol)	"Danger! High Voltage" Symbol Indicating dangerous voltage
(microwave symbol)	"Caution! Microwave" Symbol Indicating strong electromagnetic field
(laser beam symbol)	"Caution! Laser Beam" Symbol Indicating intense laser beam
(hot symbol)	"Caution! Hot" Symbol Indicating that the surface temperature of the equipment may cause scalding. Caution: Do not touch casually to avoid scalding

2) Toxic and dangerous obejects

(1) Beryllium oxide. Beryllium oxide is inevitably used in some components of the equipment. Only when the components containing beryllium oxide are damaged, can beryllium oxide in those components cause harm to human body. Personnel who contact or deal with the device should understand the characteristics of the device and take appropriate preventive measures.

Components containing beryllium oxide are used in many parts of the equipment, such as power amplifier circuit, combiner circuit and so on. Any device or module that contains beryllium oxide should not be placed in an environment where beryllium oxide can be released into the air due to mechanical damage.

Components containing beryllium oxide should not be discarded at will. Rather, they should be delt with through chemical treatment or as the special waste in accordance with local regulations.

(2) Hydrochloride. Chemicals containing hydrochlorides are inevitably used in certain components of the equipment, which will give off toxic gases when burned. Therefore, burning devices should be avoided and precautions taken to avoid people from inhaling toxic gases.

Components containing hydrochloride are not allowed to be discarded at will but should be processed through chemical treatment or be delt with as the special waste in accordance with local regulations.

(3) Hydrofluoride. Chemicals containing hydrofluoride are inevitably used in certain components of the equipment, which will give off toxic gases when burned. Therefore, burning devices should be avoided and precautions taken to avoid inhaling toxic gases. Components containing hydrofluoride are not allowed to be discarded at will and should be delt with through chemical treatment or as the special waste in accordance with local regulations.

3) Electrical safety

(1) High voltage. No electrical components involving high voltage are used in the equipment.

(2) Power cord. It is strictly forbidden to install or dismantle the power cord while it is connected to the power. Before installing or dismantling the power cord, the power must be cut off. Before cables are connected, it is necessary to confirm whether the cables and their labels are correctly matched when they are installed.

Electric sparks or arcs may be generated when the power cord contacts the conductor, which may cause fire or injury to personnel.

(3) Tools. When a person is handling power supply and alternating current, special tools must be used. Ordinary tools or those carried without consent are prohibited.

(4) Drilling. It is strictly forbidden to drill holes on the cabinet without approval. Drilling holes that fail to meet the requirements will damage the cables and devices inside the cabinet, as well as damage the appearance and shielding effectiveness of the cabinet. Metal filings produced by drilling are also prone to causing short circuit on circuit board when they fall into the cabinet. If it is highly necessary to drill holes in the cabinet, the consent from the equipment manufacturer must be obtained, and the following items must be strictly observed:

① Before drilling holes in the cabinet, one should wear insulating gloves to remove the cables inside the cabinet, and all the boards in the cabinet.

② When drilling holes, eye protection is necessary. Splashing metal filings may hurt operator's eyes.

③ Make sure that metal chippings will not fall into the cabinet.

④ Substandard drilling will damage the electromagnetic shielding performance of the cabinet.

⑤ After drilling holes, it is necessary to clean metal filings in time, and conduct rust-proof treatment around the holes.

(5) Thunderstorm. It is strictly forbidden to deal with high voltage, alternating current, or work on tower and mast during thunderstorm. In thunderstorm, strong electromagnetic field will be generated in the atmosphere. Therefore, in order to avoid the damage of equipment by lightning, lightning protection and grounding of the equipment should be carried out in time.

Keep a safe distance from the cabinet in thunderstorm. Especially do not touch the top of the cabinet and grounding wire. Maintaining the equipment during thunderstorm is strictly prohibited.

(6) Static electricity. Static electricity produced by human body can damage electrostatic-sensitive components on circuit boards, such as large-scale integrated circuits (IC). In dry conditions, static electricity on human body can reach as high as 38kV, which can be stored on human body for a long time. When the operator with static electricity contacts the device, the static electricity is discharged through the device, damaging the device itself.

In order to avoid electrostatic damage to sensitive components, it is necessary to wear anti-static wristband and ground the other end of anti-static wristband before contacting the equipment, and carrying socket board, circuit board as well as IC chip by hand.

The connection line between the wristband and the grounding point must be connected in series with a resistance of more than 1MΩ, to protect the personnel from accidental electric shock. Anti-static wristband should be inspected regularly. It is strictly forbidden to replace the cable on the anti-static wristband with other cables.

Static-sensitive boards or modules should not be in contact with objects with static electricity or prone to static electricity. For example, friction with packaging bags, transfer boxes and conveyor belts made of insulating materials will make the device itself electrostatically charged, and electrostatic discharge will occur when the device with static electricity contacts the human body or the ground, thus causing damage to the device.

Boards or modules sensitive to static electricity can only contact high-quality electric discharge materials, such as anti-static packaging bags. Anti-static bags should be used during the storage and transportation of boards.

Before the measuring equipment is connected to the board or module, static electricity of the measuring equipment should be released, which means the measuring equipment should be grounded first.

The boards or modules cannot be placed near the strong DC magnetic field, such as the cathode ray tube of the display. The safe distance should be at least 10cm.

The damage caused by static electricity can be cumulative. When the damage is slight, the device can still work, but with the damage increasing, chances are that device will suddenly fail. Therefore, there are two kinds of damage caused by electro-static discharge to the device, namely obvious one and hidden one. The hidden damage is not visible at the scene, but it could make the device more vulnerable and easily damaged under overvoltage, high temperature and other conditions.

(7) Power label. Make sure the label is correct before connecting the cables.

(8) Grounding. Operation and maintenance personnel must connect the protective grounding terminal of the equipment's housing to the earth before connecting the input power supply.

(9) Flammable air. The equipment should not be placed in an environment where flammable, explosive gas or smoke exists. Any operations should not be conducted in

environment mentioned above either. Any operations involving electricity in the aforementioned environment may be dangerous.

4) Battery

(1) In general situation. Substandard operation of batteries may cause danger. It is necessary to prevent short circuit of batteries, or the overflow and loss of electrolyte during the process. The overflow of electrolyte will pose a potential threat to the equipment, which might corrode metal objects and circuit boards, causing damage to the equipment or short circuit.

⚠Caution:

Before using batteries, please read very carefully safety instructions on battery handling and correct ways to connect them.

In order to ensure safety, before the installation and maintenance of batteries, the following aspects should be observed:

① Be careful when handling the batteries to avoid violent shaking.

② It is strictly forbidden to wear watches, bracelets, rings and other objects that contain metal.

③ Use special insulation tools.

④ Use eye protection devices, and take preventive measures.

⑤ Wear rubber gloves and aprons that can prevent the electrolyte.

⑥ The electrodes of batteries should face upward whenever they are carried. It is strictly prohibited to invert or tilt the batteries.

(2) Short circuit. The short circuit of batteries caused by metal objects should be avoided, such as the short circuit caused by improper use of tools. If permitted, one should cut off power supply of the batteries first before moving on to other steps.

⚠Caution:

Short circuit of the batteries can cause injury. Although the voltage of common batteries is relatively low, the instantaneous strong current caused by short circuit may release huge amount of energy.

(3) Noxious gas. Batteries in operation might release flammable gases. The places where batteries are placed should be kept ventilated and fire prevention measures should be taken.

⚠Caution:

Unsealed lead-acid batteries should be prohibited. The gas released from the batteries might cause burning or corrosion to the equipment. Batteries should be fixed and placed horizontally.

(4) High temperature. When the battery temperature exceeds 60℃, check acid spillage.

⚠Caution:

High temperature of the battery may cause deformation, damage to the battery and a leakage of acidic liquid.

(5) Acidic liquid. When removing or carrying the leaking batteries, possible injuries caused by the acidic liquid should be noted. When the acidic liquid leaks, it should be disposed in time. Sodium bicarbonate and sodium carbonate can be used to neutralize and absorb the acidic liquid.

⚠Caution:

When acidic liquid leaks, it should be neutralized and absorbed in time.

(6) Battery replacement. When replacing batteries, it is necessary to use the same type of battery, and take anti-electric shock measures to prevent electric shock, short circuit and other accidents.

⚠Caution:

Operation and maintenance personnel should not replace the stipulated types of batteries with other types; otherwise, there is a risk of explosion.

5) Microwave and magnetic field

When in operation, EMB5116 TD-LTE will generate certain electromagnetic radiation, so the equipment can only be installed and maintained by trained personnel. The radiation of the equipment should conform to the national standard "GB 9175—1988", which stipulates an accepted amount for electromagnetic radiation.

The standard for the accepted radiation intensity of the electromagnetic wave is divided into two levels:

Level one refers to a safe range, in which people who live and work for a long period will not be harmed.

Level two refers to a middle range, in which people who live and work for a long period might potentially have negative reactions.

As for the EMB5116 TD-LTE in TD-LTE system, when it is in operation, the accepted radiation intensity of the electromagnetic wave meets the standard required in level one, which stipulates that the energy density of electromagnetic radiation should be less than or equal to 10 $\mu W/cm^2$. In this circumstance, people living and working for a long period will not be harmed.

6) Laser

The laser beam in the optical fiber interface will damage eyes. Therefore, the equipment can only be installed and maintained by trained personnel.

⚠Caution:

The laser beam in the optical fiber interface will damage eyes. When optical fibers are installed and maintained, it is strictly forbidden to look directly into or to keep eyes close to the optical fibers, optical modules on the board, or the interface of the lightware terminal equipment.

7) High temperature

When the equipment runs normally in tropical climate, the standard temperature of the equipment is 45℃. When the equipment runs normally, the highest temperature should not exceed 75℃. When in malfunction, the highest temperature of the equipment should not exceed 100℃.

⚠Caution:

High temperature can inevitably occur to some parts of the equipment, so do not touch them at will to avoid scalding.

8) Cooling fan

When disassembling and assembling the cooling fan in operation, do not put fingers or tools into the fan before the fan is powered off and stops rotating, to avoid damage or injury to the equipment and people. When replacing the relevant parts, pay attention to conponents, screws, tools and other objects, and make sure they do not fall into the rotating fan, to avoid damaging the fan or the equipment.

9) Other objects

When plugging the boards, do not use too much strength, which might distort the pins on the backplane. Plug the boards along the slots to prevent surfaces of the circuit boards from contacting each other, which might cause short circuit.

Do not touch the circuit of the board, components, wire connectors, and wiring slots when holding the board.

Do not maintain or adjust the equipment independently without the assistance from the professionals.

[Requirements of the training]

1. Preparation of tools, meters and equipment

A set of mobile communication equipment and tools.

2. Knowledge evaluation points

(1) Composition of mobile communication equipment system.

(2) The function and role of each component.

3. Skills assessment points

(1) Draw the structural diagram of the mobile communication system.

(2) Master the performance and function of main equipment.

(3) Master the installation standard.

Task 1　Introduction to Base Station Equipment

※ Related knowledge

A complete TD-LTE mobile communication network consists of terminal equipment, base station equipment, core network equipment, network management center, transmission network equipment, and so on. Together, these devices carry various kinds of services of TD-LTE, as shown in Figure 1-1-1.

Base station equipment is the most important wireless device in the whole network.

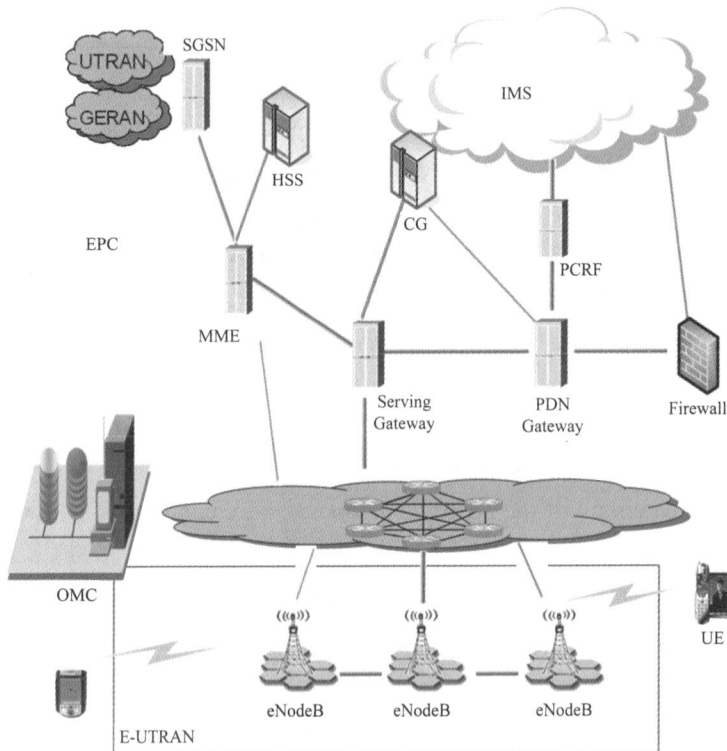

Figure 1-1-1　Network Architecture of TD-LTE

※ Practice

1. Preparation before the task

Prepare a RRU(Remote Radio Unit) device of the model EMB5116 according to the content of the practice.

2. What to learn in this task

1) Oreintation of base station products

The TD-LTE base station made by Datang Telecom Group is categorized into indoor base station and outdoor base station, each providing solutions of wireless access network for TD-LTE system according to different needs and actual situations.

EMB5116 is a kind of base-band remote base station developed by Datang Telecom Group. This technology can support both local coverage and remote coverage, thus solving siting issues.

EMB5116 can be applied to outdoor macro coverage, such as urban hot spots, suburbs, towns, rural areas, and highways, etc. The technology can quickly be applied to major business areas with a lower cost.

EMB5116 can also be applied to small and medium-sized indoor coverage, such as tunnels, subway stations, buildings, and residential areas, etc., which can improve the network coverage and service quality without significantly increasing the cost.

2) Technical characteristics of base station products

(1) Major technical characteristics.

① They are TD-LTE oriented, with an advanced architecture.

② The resource pooling is adopted to improve the utilization of hardware resources and the fault tolerance of the system.

③ Digital intermediate-frequency technology is applied to improve signal processing ability.

④ Processing capability of the sectors are strong, which can support high power coverage and wide bandwidth coverage for a single sector.

⑤ The cooling fans are intelligently controlled, which can lengthen the lifespan of the fans and reduce noise.

⑥ They support the interference suppression through in-band adaptive filtering.

⑦ They support cascaded RRUs and can flexibly expand wireless coverage.

(2) Wide coverage.

① The maximum radius of local coverage is 100km.

② With smart antenna, the sensitivity of uplink reception and downlink coverage are improved.

③ The standard single-stage distance of the fiber-based remote base station is 2km, and the single-stage maximum capacity is 10km. The multi-stage capacity can reach as long as 40km, with 4 cascades at most.

(3) Flexible configuration.

① The number of activated users supported by 20MB bandwidth in each cell can reach 400, and the number of connected users is 1,200.

② A standard configuration is 3 cells, with a maximum processing capacity of 60MB bandwidth, which can support 1,200 activated users, and 3,600 connected users.

③ The capacity expansion can be achieved by adding baseband boards.

④ O1 and O2 cell configuration are available.

⑤ S1/1/1, S2/2/2 cell configuration are available.

(4) Flexible networking.

① S1/X2 networking interface: When fully configured, a single cabinet can support two sets of electric/optical port, and FE/GE adaptive port (two ports can be solely used as electric ports or optical ports respectively, or an electric ports plus optical ports) or their combination.

② Networking mode: It supports co-frequency networking; It supports star-shaped, chain-shaped and ring-shaped networking for S1/X2 port; It also supports star-shaped, chain-shaped and ring-shaped networking for Ir port.

③ Clock source: It supports GPS synchronization, Beidou satellite synchronization, GPS/Beidou satellite optical fiber remote synchronization and upper eNode B synchronization.

(5) Flexible and convenient for installation. Small in size and light in weight, EMB5116 TD-LTE can be easily installed on the indoor wall of a building, in a 19-inch standard cabinet, or directly in the indoor environment, with no machine room and air conditioning required so that the station construction can be established rapidly with a lower cost.

(6) Easy for upgrade and expansion.

① Designed for compatibility. The board of EMB5116 TD-LTE is fully compatible with the EMB-TD series base stations of Datang Mobile Communication Equipment Co., Ltd.

② Flexible configuration. It supports omni-directional cell configuration or multi-sector configuration, which guarantees flexible coverage.

③ Smooth expansion. The capacity of EMB5116 TD-LTE can be expanded by adding baseband boards, with a maximum of 120 MB bandwidth expansion according to users' needs.

(7) Powerful operation and maintenance functions.

Mobile network management system provides management functions for operation and maintenance terminal LMT (which is different from TD's LMT-B): system status monitoring, data configuration, alarm processing, safety management, equipment operation, software configuration, monitoring management, self-configuration and self-optimizing, and tracking management, etc.

(8) Serialization of products. EMB5116 TD-LTE can support the serialized design of related functional units.

Serialization of the antenna: It supports indoor distributed antenna, and other omni-directional, dual-polarized antenna like 2-antenna form and 8-antenna form, as well as sector array antennas.

Serialization of the RRU power: It supports multi-level power amplifier.

Serialization of the bandwidth: Each baseband board supports 10MB, and 20MB bandwidth.

3) Main equipment of base station

The dimensions of EMB5116 TD-LTE's Chassis: 88 mm×483 mm×310 mm (height×width ×depth). When fully loaded, the chassis is 10kg in weight, and its height is 2 Units. The appearance is shown in Figure 1-1-2.

Figure 1-1-2   Exterior of Chassis

4) The hardware architecture of the base station's main equipment

The main equipment of EMB5116 TD-LTE includes Switching Control and Transmission Element (SCTE), Baseband Processing Operation and Ir Interface Unit (BPOX), as well as Compact Backplane (CBP), Fan Control (FC), Environment Monitor Board A/D type (EMA/EMD), Power Supply A/C (PSA/PSC) and Extended Transmission Processing Element (ETPE).

The layout of hardware units is shown in Figure 1-1-3. The fully configured position is shown in Figure 1-1-4.

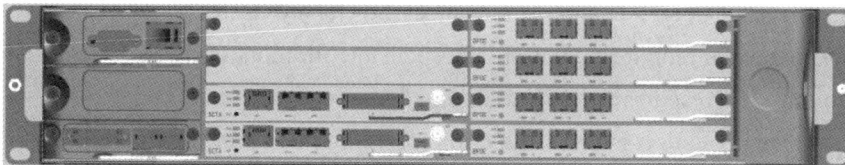

Figure 1-1-3   Diagram of the Layout of Hardware Units

SLOT 11	BPOG	SLOT 3	BPOG	SLOT 7	
PSC	BPOG	SLOT 2	BPOG	SLOT 6	
SLOT 10	SCTE	SLOT 1	BPOG	SLOT 5	FC
EMA SLOT 9		SLOT 0	BPOG	SLOT 4	SLOT 8

(a) Direct Current

PSC SLOT 11	BPOG	SLOT 3	BPOG	SLOT 7	
	BPOG	SLOT 2	BPOG	SLOT 6	
SLOT 10	SCTE	SLOT 1	BPOG	SLOT 5	FC
EMA SLOT 9		SLOT 0	BPOG	SLOT 4	SLOT 8

(b) Alternating Current

Figure 1-1-4   Diagram of a Fully Configured Board

(1) Switching control and transmission unit. Switching control and transmission unit consists of SCTE board whose major functions are as follows:

① The S1/X2 interface between EMB5116 TD-LTE and EPC can support two electric/optical ports, and FE/GE adaptive ports (two ports can be solely used as electric ports or optical ports respectively, or electric ports plus optical ports) or their combination.

② The function of service and signaling exchange.

③ All control and uplink interface protocol control-plane processing.

④ The function of high stability clock and holding.

⑤ Power-on and power-saving control of the board.

⑥ In-position detection and survival detection of the board.

⑦ Clock and synchronous stream distribution.

⑧ Frame management independent from board software.

⑨ Primary and secondary redundancy backup of the system.

(2) Baseband processing unit. Baseband processing unit consists of BPOG (Baseband Processing Operation Type-G) board. The main functions of the BPOG board are as follows:

① It implements the standard Ir interface.

② It realizes the convergence and distribution of baseband data.

③ It realizes physical layer algorithm for TD-LTE.

④ It realizes L2 functions such as MAC/RLC/PDCP of TD-LTE.

⑤ It realizes the operation and maintenance of the board itself.

(3) Environmental monitoring unit. It is responsible for monitoring the environment.

(4) Fan unit. It is responsible for heat dissipation of equipment.

(5) Power supply unit. It is responsible for providing power for the equipment.

(6) Extension transmission processing unit. It is responsible for the extension and transmission for the equipment.

5) Remote radio unit

The base station RRU contains two-channel RRU and eight-channel RRU.

(1) Exterior of two-channel RRU. It contains TDRU331FAE, TDRU332FA, TDRU341FAE, TDRU342D, and TDRU342E.

The dimensions of TDRU342E: 420mm×300mm×120mm (height×width×depth), with 12kg in weight and a capacity of 15L. The exterior is shown in Figure 1-1-5.

(2) Exterior of eight-channel RRU. It contains TDRU338FA, TDRU338D, and TDRU348FA.

The dimensions of TDRU338D: 439mm×356mm×140mm (height×width×depth), with 23kg in weight and a capacity of 23L. The exterior is shown in Figure 1-1-6.

Figure 1-1-5   Exterior of TDRU342E

Figure 1-1-6   Exterior of TDRU338D

3. Conclusion of the task

_____

_____

_____

_____

Task 2   Introduction to EPC

※ Related knowledge

The core network of LTE provides users with IP transmission services with high data bandwidth, low service delay, fast access speed and strong security; It supports a variety of different access technologies, enabling users to obtain uninterrupted services when moving between different access networks; It supports emergency call; It supports the routing optimization for roaming users; It provides operators with flexible control strategies and charging methods; It also provides abundant operation and maintenance means.

※ Practice

1. Preparation before the task

Prepare Evolved Packet Core equipment according to the content of this experiment.

2. What to learn in this task

Evolved Packet Core, or EPC for short, consists of network elements like MME, S-GW, P-GW, PCRF, CG, and HSS.

(1) MME (Mobility Management Entity)has the following fuctions:

① The function of access authentication. MME is equipped with the function of identifying whether UE (User Equipment) who is connected to the network has been authorized.

② The function of mobility management. MME is responsible for UE's access to EPC system, manages the mobility of mobile phones in idle mode, and tracks its location, such as maintaining TA (Transmitter Address) where UE is currently located, MME and other information. The mobility management function is mainly realized through the processes of attachment, separation and location update. These processes ensure a timely update of UE's location in related network entities as UE moves.

③ The function of session management. MME is responsible for the establishment, modification and release of EPC bearer, as well as the establishment and release of E-UTRAN (Evolved UMTS Terrestrial Radio Access Network) access network-side bearing, such as S1 connection release and bearer reconstruction of the S1-U port.

④ The function of MME Pool management. In order to better serve carriers, 3GPP proposed the concept of MME Pool. An MME pool serves multiple MMEs. Each MME stores all the MME IDs in the same MME Pool. Each eNodeB (Evolved Node B) is connected to all the MMEs in a Pool.

⑤ The function of clock synchronization. MME can set the remote server to the local time server, and make the home terminal work in the Client mode.

(2) S-GW (Serving GateWay) has the following functions:

① The function of the bearer management.

② The function of packet routing and forwarding.

③ The function of QoS (Quality of Service) control.

④ The function of GTP (GPRS Tunneling Protocol).

⑤ Charging function.

(3) P-GW (PDN GateWay) has the following functions:

① The function of managing UE's IP address.

② The function of bearer management.

③ The function of packet routing and forwarding.

④ The function of QoS (Quality of Service) control.

⑤ Security function.

⑥ GTP function;

⑦ Charging function.

(4) PCRF (Policy and Charging Rules Function): Policy Control and Charging (PCC) technology means that when the users' service data flow needs to be transmitted through the PS domain, the network element PCRF performs dynamic QoS and charging policy control on the users' application-level service data flow according to the characteristics of the service data flow, the carrier's strategy and the signing characteristics of the signed users, so as to realize an effective control and management of the carriers on users' service.

(5) CG (Charging Gateway): The Charging Gateway or CG for short, is an important network element in TD-LTE EPC. Its main function is to process the record and form a final

record file after obtaining call detail record from S-GW and P-GW. Then the final record file will be transferred to BOSS(Business Operation Support System)for processing by using FTP(File Transfer Protocol)mechanism.

(6) HSS (Home Subscriber Server)：is an important network element in LTE EPC, which contains user profile, performs user authentication and access authorization, and provides information about the users' physical location. Therefore, its functions focus mainly on users information signing and users authentication in LTE.

3. Conclusion of the task

_____

_____

_____

_____

Training Module 2　Installation of Base Station Equipment

[Brief description]

EMB5116 TD-LTE is a compact base station product developed for a variety of possible application environments. Small in size, simple in installation and flexible in confifuration, it supports wall hanging, cabinet and other ways. It can be used for indoor distributed applications and outdoor macro cell applications. With a design of full compatibility, it can provide carriers with abundant networking applications together with other TD-LTE base station products of Datang Mobile Communication Equipment Co., Ltd. This project is a typical work task designed for the practitioners who are involved in the network installation of mobile communication system for the first time. The work task mainly takes two typical base station installation methods as the training objectives. The contents include the structure of the base station system, preparations before the installation, main equipment installation, RRU installation, GPS system installation, etc. Through the practical training of this project, the trainees can meet the basic requirements for installing and adjusting the mobile network.

[Elements of the training]

1. Knowledge objectives

(1) Students master the structure of mobile communication base station.

(2) Students master the installation process of mobile communication base station.

(3) Students master the installation methods in two typical environments (indoor and outdoor).

2. Ability objectives

(1) Students can draw the structure of mobile communication base station.

(2) Students can carry out the indoor installation of mobile communication base station smoothly according to the standard.

(3) Students can carry out the outdoor installation of mobile communication base station smoothly according to the standard.

3. Quality standards

1) Installation of indoor equipment

(1) The frame and the base (brackets) should be firmly connected, and the frame must be stable after installation.

(2) The RRU cable connector should be installed in place to avoid the abnormal standing-wave ratio caused by false connection, which will affect the normal operation of the system.

(3) Trunk cable connectors should be firmly connected. As for the cable plugs made on site, they must meet the specifications, with reliable crimping and intact appearance.

(4) Make sure that the plugs of all kinds of cables must be clean before plugging, and if necessary, clean them according to the specifications.

(5) The environmental alarm acquisition line connected to the base station should be treated with reliable lightning protection measures.

(6) The surge protective devices outside the cabinet should be installed correctly and connected firmly according to the design requirements.

(7) All parts of the frame should not be deformed to affect the appearance of the equipment.

(8) The anchor bolts between cabinet or base (brackets) and ground should be installed firmly and reliably, and sequence of installing various insulating washers, plain washers, spring washers as well as bolts and nuts should be correct.

(9) After all the cabinet door panels at the main aisle side are installed, they should be in a straight line with an error of less than 5mm.

(10) The vertical deviation of the cabinet should be less than 3mm.

(11) The gap between adjacent bases should be no more than 3mm.

(12) When rigid cables are connected in the cabinet, they should not be crossed as much as possible, nor be stretched too tightly. There must be margin at the corner, and rigid cables should be neatly bent.

(13) When cables are installed in the frame, the horizontally arranged ones should be bound at the inner side of the corresponding cross bar; When vertically arranged cables are installed in the frame, they should be installed in the wire troughs on both sides.

(14) All cables in the frame should pass through the square hole on the lower part of the cabinet's side and be connected to their corresponding positions.

(15) Screws on fans should be correctly installed and firmly tightened. Make sure that fan switch, rotation and alarm are normal.

(16) The allocated insulating materials should be used for cabinet installation to keep the cabinet insulated from the ground and wall.

(17) Casing or insulating tape should be used to cover the peeled part of the cable processed on site.

(18) When the cabinets are combined, the small cover plates at the side of the cabinets must be removed, and the connecting parts on top of the cabinets should be docked correctly.

(19) Unused ports of the combiner, demultiplexer and CDU should be screwed on with matching terminals. Terminals which will not be used temporarily on the top of the cabinet should be protected with protective casing to avoid dust or grime falling into these ports.

(20) When the base level is adjusted through adjusting bolts, put a pad under the base to

increase the stress area, so as to avoid the gradual deformation of the base.

(21) The connecting parts on the top of the cabinets must be installed and fastened.

(22) The power line and signal line of the alarm box installed on the wall should be routed through PVC trough, and the redundant wires should be well-coiled and placed under the floor at the side of the alarm box or on the wiring rack.

(23) When cables are bound, they should be bound neatly in appearance and moderately in tightness, with buckles facing the same direction. Redundant buckles should be cut off, and all the buckles must be trimmed so that there will be no spikes.

(24) The installation position of the cabinet should conform to the engineering design; the cable routing should conform to the engineering and design documents, which will be convenient for maintenance and future expansion.

(25) The upper surface of the anti-static floor laid in the machine room should not be higher than the lower surface of the base (so as not to affect the heat dissipation effect of the base).

(26) Make sure that the active accessories of the cabinet act normally.

(27) Fill in tags according to specifications and then paste them neatly to cables, with all of them facing the same direction. They can be made according to the customers' requirements to make inspection easy. Generally, it is recommended that the tag be pasted 2cm away from the plug.

(28) The connecting cables from the combiner, demultiplexer, CDU to the top of the cabinet should pass through the gaps on both sides of the cabinet.

(29) All reserved handles and panels should be installed.

(30) The board can be unplugged and plugged smoothly. If there are screws on the board's panel, make sure that they should be properly tightened and the spring steel wires are intact.

(31) Jumpers should be arranged according to layers and sectors.

(32) The alarm box should be installed correctly according to the design requirements, and the panel should be clean without obvious damage.

(33) When installing components of the frame, please avoid paint-falling or collision, which may damage the appearance of the equipment. In that case, repair the fallen paint.

(34) Appropriate margin should be reserved at the cable plug.

2) The antenna feed of base station

(1) All outdoor jumpers and feed line joints should be sealed and go through waterproof treatment according to the specifications.

(2) The antenna should be placed in the protection area of the lightning rod.

(3) There is no obvious folding and twisting on the feed line, and no exposed copper sheet.

(4) The feed line connector should be made in a standard way without looseness.

(5) The connection between antenna stand and tower should be reliable and firm, and the connection between antenna and antenna stand should also be reliable and firm.

(6) The main feed line is connected correctly and the sector is correct.

(7) The omni-directional antenna should be kept vertical, and the error should be less than ±2°.

(8) The azimuth error for the directional antenna should not be greater than ± 5°, and the dip angle error for the directional antenna should not be greater than ± 0.5°.

(9) When the feed line is bent downward from the roof over the external wall, the contact part with the wall corner should be protected with casing. When the feed line enters the room along the wall from the roof, fixing clips should be used to fix the feed line to avoid swinging.

(10) Make sure that antenna lightning arresters (on upper and lower part) of GPS is correctly installed. The protected end of indoor lightning arrester should face the base station, and the protected end of outdoor lightning arrester should face the GPS antenna (check this if there is GPS antenna).

(11) When the feed line window on the roof is installed so that the feed line can enter the room, make sure that the window is sealed properly, and the jumper should be provided with an elbow pipe to avoid water at a section where incoming feed line and antenna' outgoing line run.

(12) For GSM macro base station, it is required that the lightning arrester go through lightning protection and grounding treatment according to the standard, and the grounding wires, after being converged, should be connected to the outdoor protective ground bus bar of the machine room.

(13) Fixing clips on feed lines should be evenly spaced, with all clips facing the same direction, and the fixing clips should also be firmly installed.

(14) Feed lines are forbidden to be laid across. Rather, it is required that the arrangement of incoming feed lines should be neat and straight, horizontally or vertically, and with the same curvature.

(15) The installation position of the antenna should be consistent with the design documents.

(16) The distance between omni-directional antenna and tower should be no less than 1.5m; the distance between directional antenna and tower should be no less than 1m.

(17) Requirements for horizontal distance between main and diversity antenna of omni-directional antenna: For M900, the distance should be no less than 4 m; For M1800, the distance should be no less than 2m.

(18) Requirements for the horizontal distance of the main and diversity antenna of the monopole directional antenna: For M900, the distance should be no less than 4m; For M1800, the distance should be no less than 2m.

(19) The vertical distance between two directional antennas installed on the same antenna stand should be no less than 0.5m.

(20) The top of the casing of the omnidirectional antenna should be flush with or slightly higher than the top of the stand.

(21) When tower mounted amplifier is installed, the side connected to the antenna should face up, the side connected to the feed line should face down, and the tower mounted amplifier

should be installed near the antenna.

(22) Paste and bind tags of communication cables, feed lines and jumpers according to the specifications. The tags should be neatly arranged and face the same direction.

(23) The minimum bending radius of the feed line should be no less than 20 times the diameter of the feed line.

(24) The GPS antenna stand should be stably installed, with no blockage within its vertical angle of 90 degrees (check if there is a GPS antenna).

(25) When the omni-directional antenna is installed on the roof, the blind area should be avoided as much as possible.

(26) The jumper connected to the antenna should be tied to the tower steel frame along the cross bar.

(27) There should be margin for bindings when their redundant bands are cut.

(28) When the front of the cabinet is parallel to the direction of incoming feed line, or the back of the cabinet is directly facing the direction of incoming feed line, sectors should be arranged in horizaontal rows, with the same arrangement sequence in each row; when the front of the cabinet is directly facing the direction of incoming feed line, sectors should be arranged in vertical rows, with the same arrangement sequence in each row.

(29) When a feed line is routed in, the indoor and outdoor part should be kept straight, with a gap of over 0.5m.

3) Power line and grounding

(1) Copperdad cables and the whole section of material should be used in all power lines and grounding wires, with no joints in the middle.

(2) When cable lugs at both ends of power line and grounding wire are made, they should be welded or crimped firmly.

(3) The power line and grounding wire outside the cabinet should be bound separately from other cables.

(4) The grounding wire and power line of the equipment should be connected correctly. The correct installation sequence of the plain washers, spring washers and nuts is required. Tighten the nuts in place.

(5) The redundant length of the grounding wire and power line should be cut off and should not be coiled.

(6) DDF (Digital Distribution Frame) and base station environmental alarm box must be reliably grounded as required.

(7) For the grounding of antenna feed line on the top of a building without tower, it should be connected to the rooftop lightning protection network nearby.

(8) The power line, grounding wire cable lugs, and bare wires should be wrapped with casing or insulating tape, with no copper wire exposed.

(9) For the grounding terminal, rust removal and cleaning treatment should be carried out

before connection, so that reliable connection can be guaranteed; after the connection, anti-corrosion and anti-rust treatment should be carried out for the grounding terminal with fast-dry spray paint, to ensure a good performance of the grounding terminal over a long period. Each terminal should be correctly and reliably installed with plain washers and spring washers.

(10) A reliable grounding point should be provided for GPS antenna feed line; the lightning arrester of GPS antenna feed line does not need to be grounded indoors or outdoors, and lightning protection and grounding can be done near the outdoor GPS antenna. (check this if there is GPS antenna).

(11) The front door of the cabinet should be connected to grounding bolts on the fixed door plate by using cables distributed by the manufacturer or purchased from the manufacturer. The connecting grounding wire of the cabinet door plate should be connected firmly and reliably.

(12) The grounding clips of feed line should be fixed well and directly on the steel plate of the tower nearby. The grounding clips of feed line should be made according to specifications, and they should be firmly connected and treated with rust prevention, corrosion prevention and water proofing.

(13) When the tower is higher than 60m, another feed line grounding clip should be added in the middle of the tower.

(14) If the feed line leaves the tower and continues to run for some distance on the top of the building (or on the wiring rack) before it enters the room, and if the distance exceeds 20m, then a lightning protection grounding clip should be added on the top of the building (or on the wiring rack).

(15) The distance between power line, grounding wire and signal line outside the cabinet should be kept more than 3cm away.

(16) The feed line should be grounded at three points at least, from the top of the tower to the machine room (within 1m after it leaves the platform of the tower; within one meter before it leaves the tower and runs on outdoor wiring rack; within 1m from the feed line window), and the grounding points should be bound firmly, with good waterproof treatment.

(17) The routing of power line and grounding cable should be consistent with the engineering design documents, which will be convenient for maintenance and expansion.

(18) When the feed line is routed into the room along the wall from the top of the building, and if the down-lead ladder is used, the ladder should be grounded.

(19) The grounding wire of feed line should be directed from top to bottom; the angle between the feed line and the grounding wire should be kept at no more than 15°.

(20) The diameter of the equipment power line, grounding wire and equipotential bonding wires between cabinets should meet the power distribution requirements.

(21) Make sure that equipment power switch, cooling fan and other functions are normal.

(22) Tags should be filled in and be pasted on the power line and grounding wire according to the specifications. Tags should be neat and face the same direction (including the

tag of power distribution switch). They can be made according to the customer's requirements, which will be convenient to inspect. Generally, it is recommended the tag be pasted 2cm away from the plug.

(23) The distance between power line, grounding wire and signal line outside the cabinet should be greater than 3cm.

(24) When two or more wires and cables are installed on a binding post, generally, they should be installed in a cross or back-to-back manner. When they overlap, 450 type or 900 type bend on the cable lug is recommended. In case of overlapping installation, the larger cable lug should be installed at the lower part and the smaller cable lug should be installed at the upper part.

(25) The power supply and grounding wire should be tied straightly and neatly, and smooth at the bending.

(26) Power line and grounding wire distributed by the manufacturer: –48V power line adopts blue cable; GND (BGND) grounding wire adopts black (red) cable; PGND protective grounding wire adopts yellow green or yellow cable.

4) Other standards

(1) Check and test the power-on hardware of the equipment.

(2) Other items which are not mentioned above but are under assessment requirements, refer to the engineering design documents and the installation manual.

(3) Installation requirements should be negotiated with customers before construction.

(4) Tools used during installation should meet the anti-static requirements.

(5) The expansion signal cable should be bound or plugged to the reserved position inside the cabinet to be expanded, so as to facilitate future expansion maintenance and to avoid loss.

(6) The unused plugs should be protected with protective caps, etc.

(7) Unpacking, plugging and unplugging process of the board should conform to anti-static operation standard.

(8) Gaps on the floor around the cabinet frame should be filled up, and there should be no cut buckles, screws and other objects under the floor.

(9) The machine room should be clean and tidy, with all waste items like packing boxes removed. The remaining items after the installation should be stacked orderly and properly.

5) Environment for equipment installation

(1) The DC power supply voltage and capacity provided by the machine room should meet long-term safe operation requirements of the equipment.

(2) PDF (junction box) and primary power output current limiting fuses should meet the requirements for the operation.

(3) The grounding resistance of base station should be less than 5Ω, and the grounding resistance of base station in areas where annual thunderstorm days are less than 20 days should be less than 10Ω.

(4) The outdoor grounding copper busbar should be led to the underground grounding grid through a dedicated and reliable path, and the routing of grounding wires should meet the design specifications.

(5) The E1 line should not be hung overhead the base station. If the E1 line is to be hung overhead, safety protection measures should be taken for the line.

(6) The anti-seismic, lightning protection and bearing capacity of the machine room should meet the construction requirements and the long-term safety requirements.

(7) Relevant parameters of AC power supply should meet the requirements for long-term safety operations. The AC power supply system in the machine room should be equipped with lightning protection unit which should be reliably grounded.

(8) The ambient temperature and relative humidity in the machine room should meet the requirements for long-term safety operations.

(9) The machine room should be equipped with appropriate fire prevention measures.

(10) It is recommended that protective grounding treatment be provided for metal structures which are connected to the equipment in the machine room.

(11) The machine room should be far away from areas with strong electric field, strong magnetic field, strong electric wave, or strong heat source, etc. The machine room should meet the requirements for electromagnetism.

(12) Low voltage power cable should be buried and led to mobile base station.

(13) When the distribution transformer is in the mobile communication station, the grounding grid of the machine room should be connected with the grounding grid of the distribution transformer to form a unified grounding grid.

(14) In the mobile communication station, the grounding grid of tower and the grounding grid of machine room should form a unified grounding grid.

(15) A set of grounding body should be shared by the working grounding, protection grounding and lightning protection grounding of the building in the mobile communication station.

[Requirements of the training]

1. Knowledge evaluation points

(1) Students should be clear about working hours and technologies as required by the installation task list of base station equipment, and assign tasks to the corresponding personnel.

(2) Students can correctly identify and install all base station equipment, and correctly complete the routing and connection of various cables.

2. Skills assessment points

(1) Students can install base station equipment according to drawings, technological requirements and safety regulations.

(2) After the construction, students can carry out the test operation by using OEM software according to the task list.

Task 1　Installation of Main Equipment System

※ Related knowledge

The main equipment of TD-LTE base station contains one EMB5116 host, one RRU, one set of GPS antenna and feed line, as well as power supply, lightning protection equipment, waterproof equipment, etc. The system connection diagram is shown in Figure 2-1-1.

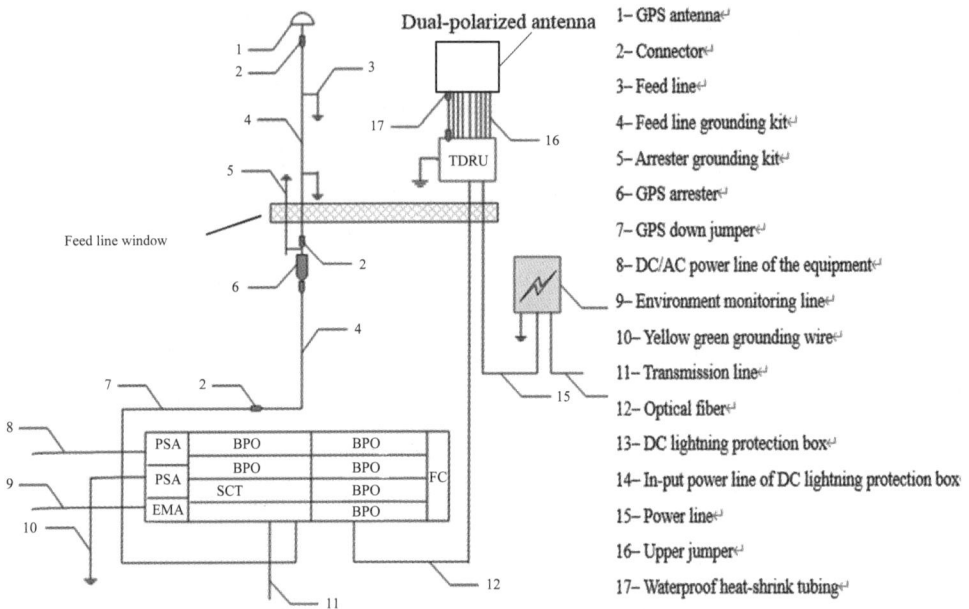

1– GPS antenna
2– Connector
3– Feed line
4– Feed line grounding kit
5– Arrester grounding kit
6– GPS arrester
7– GPS down jumper
8– DC/AC power line of the equipment
9– Environment monitoring line
10– Yellow green grounding wire
11– Transmission line
12– Optical fiber
13– DC lightning protection box
14– In-put power line of DC lightning protection box
15– Power line
16– Upper jumper
17– Waterproof heat-shrink tubing

Figure 2-1-1　System Connection Diagram (outdoor)

※ Practice

1. Preparation before the task

1) Inspection and verification of the installation environment

In order to ensure a safe, stable and reliable operation of EMB5116 TD-LTE compact base station, air conditioning and dehumidification heater should be installed in the equipment room.

The floor of the machine room should be paved with anti-static floor, under which is the concrete foundation (there should be no gap between the anti-static floor and the concrete foundation). The mark number for the concrete is required to be greater than 250, which can then firmly fix steel anchor bolts.

If the main equipment is wall-mounted, the wall shall be cement wall or brick (not air brick) wall. Use four M6 anchor bolts to fix the frame and then install the main equipment.

Good lighting should be provided in the machine room, and 220V power sockets (3-pin) should be installed on the walls around

The walls and ceiling in the machine room should be painted and clean.

The wiring rack on which cables are laid and the outlet should be installed and constructed as required before the equipment is installed in the machine room.

(1) The resistance value of equipment grounding wire in the machine room should be less than 5Ω. The grounding bus bar should be set under the wiring rack near the bottom of the wall of the machine room, so as to facilitate the installation of the equipment grounding wires.

(2) Other environmental conditions should meet the relevant requirements listed in Part 4 of GB/T 4798.4—2007 about the environmental conditions for the application of electrical and electronic products, which stipulates the standards for places without climate protection.

The EMB5116 TD-LTE is indoor equipment, therefore the temperature and humidity requirements are as follows:

① Ambient temperature: –5~+55℃(long term); –20 ~ +70℃ (short term).

② Relative humidity: 15% ~ 85% (long term); 5% ~ 95% (short term).

The antenna and feed line subsystem is outdoor equipment, therefore the temperature and humidity requirements are as follows:

① Ambient temperature: –40~+55℃ (long term); –40~+70℃(short term).

② Relative humidity: 5% ~ 98% (long term); 2% ~ 100% (short term).

2) Preparation of the tools

Tools required for installation are show in Table 2-1-1.

Table 2-1-1   Tools Required for installation

##Measuring and marking tools: measuring tape, gradienter, and markers
##Drilling tools: hammer drill and its matching drill bits, and vacuum cleaner
##Fastening tools: flat-blade screwdriver, cross-head screwdriver, medium adjustable wrench, socket wrench, box-end wrench, and Allen wrench
##Bench work tools: needle-nose pliers, diagonal pliers, pincer pliers, files, hand saws, wire strippers, handle crimping pliers, wire cutters, RJ45 crystal head crimping pliers, hydraulic pliers
##Auxiliary tools: brush, medium claw hammer, paper cutter, leather dust remover, electric soldering iron, solder wire, ladder, rubber hammer, compass, torque wrench, chamferer, heat gun (electric heat gun or liquefied-gas-powered heat gun gun)
##Anti-static tools: anti-static wrist strap and anti-static gloves
##Other tools: multimeter, 500V megohmmeter (used for insulation resistance measurement), optical power meter, grounding resistance meter, antenna standing wave ratio tester, and inclinometer

2. What to do in this task

Install the main equipment of base station system.

3. Training process

The exterior of main equipment is shown in Figure 1-1-2.

There are three ways to install the main equipment chassis: 19-inch cabinet-mounted, wall-mounted, and frame-mounted.

1) Installation of main equipment chassis

(1) Chassis installed in 19-inch cabinet. There are two types of 19 inch standard cabinet which can be used to install EMB5116 TD-LTE:

① When a 19-inch standard cabinet for EMB5116 TD-LTE is available in a machine room, the room inside should be 2 Units in height and greater than 500mm in depth. In addition, 100mm should be left between the frame column of 19-inch cabinet and the front door, which are reserved for the wire arrangement. When the above requirements are met then EMB5116 can be installed, as shown in Figure 2-1-2.

② Purchase the 19-inch standard cabinet provided by Datang to install EMB5116 TD-LTE.

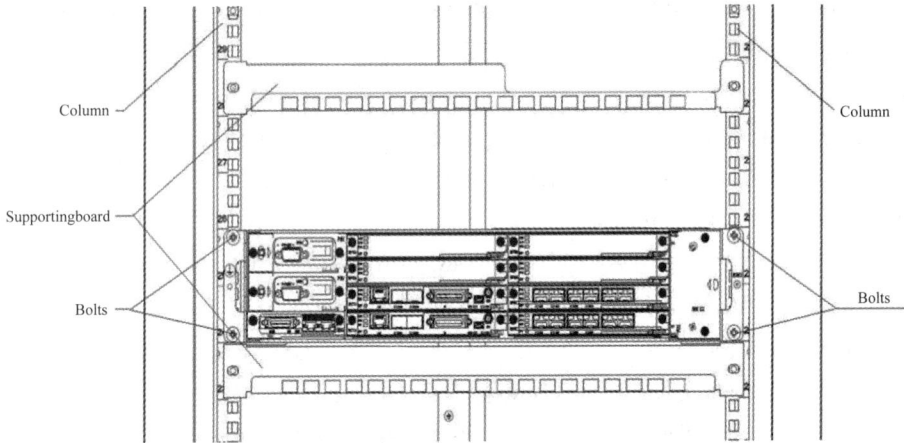

Figure 2-1-2　Installation Diagram in 19-inch Cabinet

The following installation steps will be introduced by taking the 19-inch standard cabinet provided by Datang as an example.

① Select the installation position of EMB5116 TD-LTE in the 19-inch standard cabinet. There is no special requirement for the position. Conditions permitting, a height that is convenient for installation and maintenance should be selected first.

② Remove the 2-Unit reserved panel at the selected position, then check whether the screw subassemblies fixing the panel, and the crown (floating) nuts on the cabinet column are in good condition. Replace them if there is any damage.

③ Place EMB5116 TD-LTE flat on the mounting bracket, and push EMB 5116 TD-LTE into the 19 inch cabinet with equal force, so that the mounting lugs on both sides of EMB 5116 TD-LTE can cling to the columns of the cabinet.

④ Fix EMB5116 TD-LTE on the columns of cabinet with fixing screw assemblies (4 sets in total), as shown in Figure 2-1-2.

(2) Wall-mounted chassis. When the main equipment of EMB5116 is installed on the wall, the wall should be cement wall or brick (not air brick) wall, and the thickness of the wall should

be greater than 70 mm. When mounting lugs are adopted for installation, the connection between the wall mounting components and the host is shown in figure 2-1-3. Steps are as follows:

① First, fix the left and right wall mounting components to both sides of EMB5116 TD-LTE as shown in Figure 2-1-3, and tighten the four M6 screws on both sides.

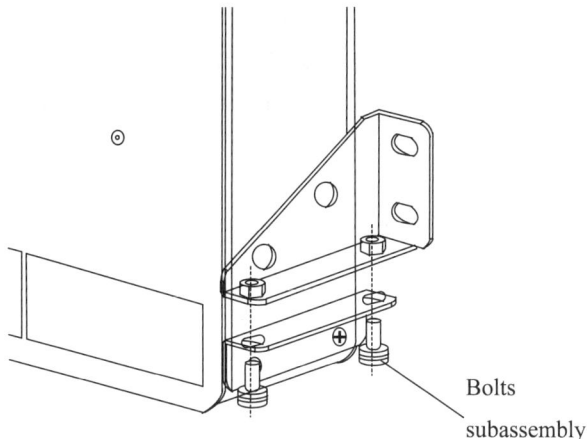

Figure 2-1-3   Diagram of Mounting Lugs Connection for EMB5116 TD-LTE

② EMB5116 TD-LTE which has already been fixed with wall mounting components should be placed horizontally at the installation position on the wall, and the drilling positions on the wall should be marked with a marker, according to the center of the installation holes on the wall mounting components; after the position of the wall mounting components is determined, two installation holes for M6 (or M8) anchor bolts should be drilled in vertical direction, and the vertical spacing between the two holes is 32mm for installing four M6×65 (or M8×65) anchor bolts; (note: from the beginning of this task, the anchor bolt is changed from M8 to M6, the length unchanged, but the stock of M8 bolts needs to be consumed, so M8 bolts will still be used in the early stage of the project).

③ Install four M6×65 (or M8×65) anchor bolts into the installation holes on the wall, making sure that the end face of the sleeve of the anchor bolt is on the same plane with the wall.

④ Mount the main equipment panel downward on the anchor bolts according to the direction shown in Figure 2-1-4.

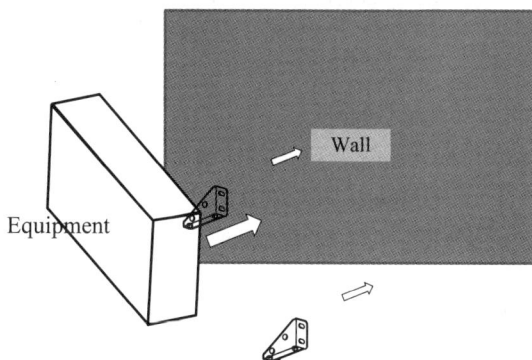

Figure 2-1-4   Diagram of EMB5116 TD-LTE Wall Mounting

⑤ Tighten the four anchor bolt subassemblies in sequence of plain washers, spring washers and nuts.

(3) Frame-mounted chassis. When EMB 5116 TD-LTE frame is installed, installation steps in three directions of the frame are the same. Therefore the steps are only decribed in the following scenario:

① Place the frame (without the main equipment of EMB5116 TD-LTE installed) horizontally at the position reserved for installation on the wall, and mark the drilling positions on the wall with a marker according to the center of the installation holes on the wall mounting components.

② Use the hammer drill to drill holes with diameter of 10mm for anchor bolts at the designated positions.

③ Install four anchor bolts into the installation holes on the wall, making sure that the end face of the sleeve of the anchor bolt is on the same plane with the wall.

④ Mount the frame on the anchor bolts according to the direction shown in Figure 2-1-5, then tighten four anchor bolt subassemblies in the sequence of plain washers, spring washers and nuts, and then screw the spring washers for half a circle after it is pressed flat.

⑤ Install the EMB5116 TD-LTE main equipment into the machine frame in the direction shown in Figure 2-1-5, and fix the main equipment to the machine frame by tightening screws.

⑥ Connect, arrange, and bind cables and wires according to the position shown in Figure 2-1-6.

Figure 2-1-5　Diagram of the Frame Installation

2) Installation of main equipment cables

The power line, transmission line, optical cable, grounding wire, 26-pin environmental monitoring line and GPS down jumper of EMB5116 TD-LTE should be connected to the front panel. EMB5116 TD-LTE needs to be connected with the following cables:

−48V DC power line or 220V AC power line, NB-RRU optical fiber, Cascaded optical fiber, grounding wire, GPS down jumper, 26-pin environmental monitoring line.

The connection position of each cable is shown in Figure 2-1-6.

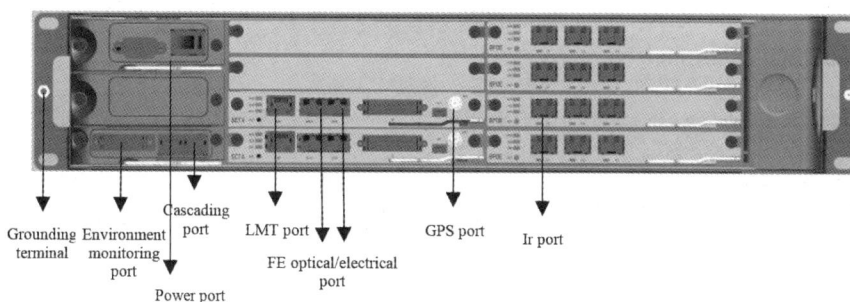

Figure 2-1-6   Diagram of the Panel Ports of EMB5116 TD-LTE

(1) Installation of main equipment grounding wire. RVVZ single-core (16 mm^2) (yellow green) wire is used as the main equipment grounding wire whose length can be cut according to the actual usage. One end should be connected to the main equipment grounding terminal, as shown in Figure 2-1-7, and the other end should be connected to the indoor grounding bar.

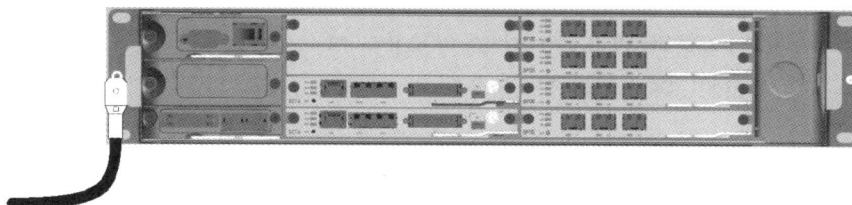

Figure 2-1-7   Diagram of Grounding Wire Connection Position on Main Equipment

Installation steps:

① Use a wallpaper knife to peel out the core wire 15mm at both ends of the cable.

② Crimp the 16mm^2-M4 and 16mm^2-M8 cable lugs onto the cores at both ends respectively with hydraulic pliers.

③ Cut two 50mm heat-shrink tubings, and use a heat gun to heat and shrink the joints between cable lugs and cables at both ends.

④ Bind the grounding wire once every 400mm with white 5mm×250mm binding tape, according to the requirements for indoor cable binding. The binding direction should be the same. Use diagonal pliers to cut off the redundant binding tape.

⑤ Identification tags should be bound to the places 50mm from both ends of the cable. Then paste the printed stickers onto the tags.

(2) Installation of main equipment power line.

① DC power line: the DC power line of the main equipment adopts a fixed length power line, with one end of D-Sub two-cored connector and the other end hanging in the air, which can be selected according to the field use.

Installation steps:

Connect D-Sub end of the power line to the power port of the main equipment, as shown in Figure 2-1-8. Tighten the screws at both ends of the connector.

Connect suspending end of the power line to the power distribution cabinet. The blue-cored wire should be connected to the –48V terminal, and the red or black cored wire should be connected to the 0V terminal.

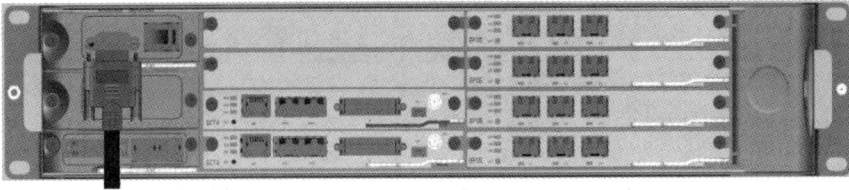

Figure 2-1-8   Diagram of the Power Line Connection Position of Main Equipment

Bind the power line once every 400mm with white 5mm×250mm binding tape, according to the requirements for indoor cable binding. The binding direction should be the same. Use diagonal pliers to cut off the redundant binding tape.

Both ends of the cable should be marked with identification tags at a distance of 50 mm from the joint. Then paste the printed stickers onto the tags.

② AC power line: the AC power line of the main equipment adopts the three-phase AC power line.

Installation steps:

Connect one end of the power line to the power port of the main equipment, as shown in figure 2-1-8.

Connect the three-phase plug to the AC power socket (if no power socket but power box for the AC power supply is provided by the operator, then the three-phase plug can be cut off, and three core wires should be crimped to the cable lugs respectively before they can be connected to the terminal of the power box).

The binding, fixing and indentifying methods and requirements for AC power line are the same as those for DC power line of main equipment.

(3) Installation of NB-RRU optical fiber. According to the design requirements, the NB-RRU optical fiber which is connected with RRU should be connected to the corresponding optical module on the front panel of the BPOG board of the main equipment, as shown in Figure 2-1-9. The port number of optical module is Ir0-Ir5.

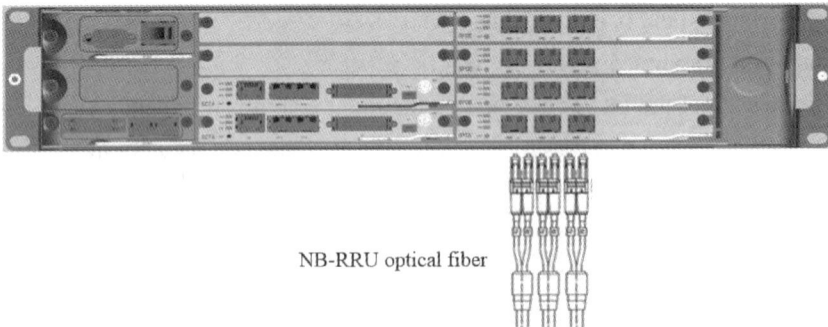

NB-RRU optical fiber

Figure 2-1-9   Diagram of NB-RRU Optical Fiber Connection of Main Equipment

Installation steps:

① Insert the DLC plug of optical fiber into the corresponding optical module of the main equipment;

② Make sector identification, and use base station aluminum signboard (A-Fiber, B-Fiber or C-Fiber) of corresponding sector at a distance of 50mm from the optical fiber output device. Use two 2.5mm×100mm (white) binding tapes to bind the signboard to the optical fiber, with the same binding direction. Use diagonal pliers to cut off the redundant binding tapes along the buckles. After the optical fiber runs out of the feed line window, use the same signboard again, and use the diagonal pliers to cut off excess binding tapes, leaving 3 to 5 buckles to spare;

③ For optical fiber binding, use white 5mm×250mm binding tape to bind optical fiber once every 400mm, according to the requirements for indoor cable binding. The binding direction should be the same. Use diagonal pliers to cut off the redundant binding tape.

(4) Installation of S1/X2 port. In EMB5116 TD-LTE, the S1/X2 port supports GE/FE, and the external ports are on the left side of the SCTE board, which are two RJ-45 ports or optical ports. Refer to Figure 2-1-10 for the position of optical fiber port.

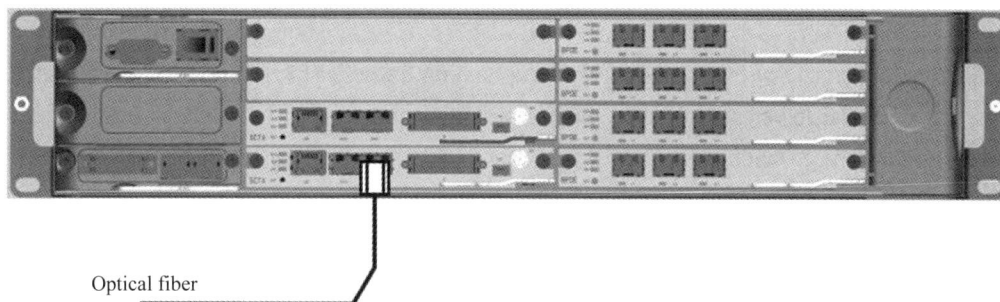

Figure 2-1-10   Diagram of Transmission Line Connection

The external basic configuration of EMB5116 supports adaptive Ethernet. The basic configuration is in the form of SFP (Small Form Pluggable) optical port. When electric connection is required, the system can realize the connection by selecting RJ-45↔SFP switching accessories.

One end should be installed on the SCTE panel and the other end should be installed on the optical port board of ODF. Each SCTE is connected to the optical transmitter and receiver through DLC single-mode fiber.

(5) Installation of environmental monitoring line. The environmental monitoring of EMB5116 TD-LTE is realized in EMA unit. When the environmental monitoring is realized by backbone node mode, it needs to use the environmental monitoring cable with SCSI 26-pin at one end to connect to the environmental monitoring port of the main equipment. Refer to Figure 2-1-11 for the connection position.

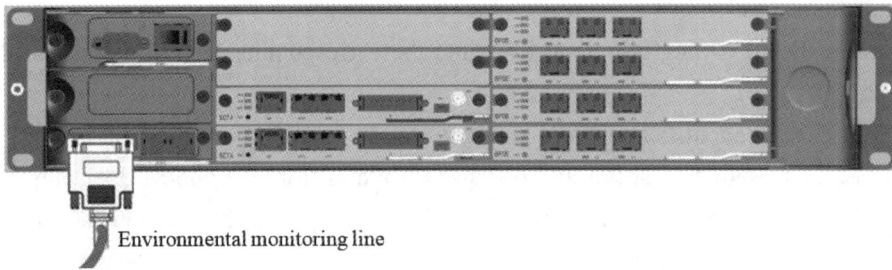

Environmental monitoring line

Figure 2-1-11   Diagram of Environmental Monitoring Line Connection

3) Specifications on the installation and arrangement of main equipment cables

(1) Specifications on the installation and wiring arrangement in the 19-inch cabinet scenario. When the main equipment is installed in a 19-inch cabinet, the connection position and wiring arrangement in the main equipment are shown in figure 2-1-12.

① Optical fiber, GPS feed line and NB-RRU optical fiber should be run through the wiring trough on the lower side of the equipment and the binding frame on the right side of the cabinet, before they are routed out from the outlet (oblong hole) on the right side of the top of the cabinet.

② The power line of the main equipment EMB5116 TD-LTE should be run through the wiring trough on the lower side of the equipment and the binding frame on the left side of the cabinet. The wiring trough at the lower side of the power distribution unit should be connected to the power distribution unit. The grounding wire of the main equipment should be run through the wiring trough on the lower side of the equipment and the binding frame on the left side of the cabinet, before it is routed out from the outlet (oblong hole) on the left side of the top of the cabinet.

(2) Specifications on the installation and wiring arrangement in the wall mounting scenario. When the main equipment is installed on the wall, the connection position and wiring arrangement of the main equipment are as follows:

① The grounding wire, power line, optical fiber, GPS feed line and NB-RRU optical fiber need to be run downward to the wiring trough on the lower side of the main equipment, and be routed out from the trough near the joint with the equipment at the other end (or grounding point). If the equipment at the other end is placed outdoor or at a great distance, wires need to be run along the wiring rack after leaving the trough.

② The gap between the upper part of the wiring trough and the subface of EMB5116 TD-LTE should be no less than 200mm.

(3) Specificationons on the installation and wiring arrangement in the frame scenario. When the main equipment frame is installed, the connection position and wiring arrangement of the main equipment are as follows:

① The grounding wire, optical fiber, GPS feed line and NB-RRU optical fiber need to be bound on the left side of the cabinet. Then the wiring trough should be used and the wires should be routed out from the trough near the joint with the equipment at the other end (or grounding point). If the equipment at the other end is placed outdoor or at a great distance, wires need to be run along the wiring rack after leaving the trough.

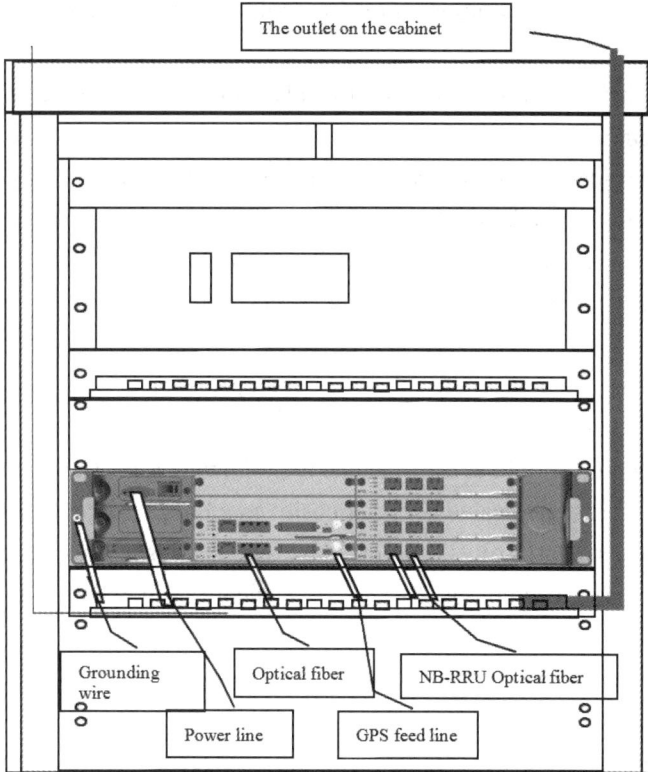

Figure 2-1-12   Installation and Wiring Arrangement in 19-inch Cabinet

② The power line of the main equipment EMB5116 should be bound on the left side of the frame, and then it should come out of the frame from the lower part of the frame, and then returns to the frame from the inner side on the right side of the frame to be bound, before it is connected to the first terminal of the DC distribution lightning protection unit.

③ After the RRU power line is routed from the terminal of DC lightning protection unit, it should be bound on the right side of the frame. Leaving the frame, it should be run through the wiring trough and wiring rack, and then connected to the power port of RRU.

④ The minimum bending radius of the cable should be considered when it comes to the distance between the wiring trough and the frame, which should be no less than 200mm.

4) Installation inspections

The inspection should be carried out after the installation is completed, and the unqualified items should be rectified.

(1) Inspection of 19-inch cabinet installation.

Cabinet inspection should at least include the following aspects:

① Check whether the horizontal, vertical and stable installation of cabinet meet requirements.

② Check whether all bolts are tightened, whether plain washers, spring washers and nuts are complete, and whether the sequence is correct.

③ Check whether the routing direction and outlet position of the cabinet meet requirements.

④ Check whether there are construction residues or tools inside the cabinet, and whether there are construction marks.

⑤ If there is any damage, deformation or peeling paint in the cabinet during construction, remedial measures should be taken in time.

⑥ Clean the cabinet, dispose the waste, and count tools.

⑦ Check whether there is collision with cables when the cabinet door is opened and closed.

⑧ Check whether the door of the cabinet can be opened and closed smoothly, and whether the door lock is locked properly.

Electrical inspections should at least include the following aspects:

① Measure the resistance value between positive and negative pole of DC circuit, and the resistance value between phases of AC circuit. Make sure there is no short circuit or open circuit.

② Check whether the color of AC line is standard, and whether the safety signs are complete.

③ Check whether the connection stability, line sequence and polarity of DC output and battery are correct.

④ Check whether the electrical components are connected and fixed firmly.

⑤ Check the corresponding relationship between DLC optical cable ports and sectors, and make sure there is no error.

⑥ Check whether all air switches are in correct state and are operated correctly during construction.

⑦ Check whether the grounding wire is connected correctly and firmly.

⑧ Check whether the wiring is neat and whether the cable binding meets requirements.

(2) Inspection of wall mounting installation.

① Check whether the horizontal, vertical and stable installation of the main equipment EMB5116 TD-LTE meet requirements.

② Check whether all bolts are tightened, whether plain washers, spring washers and nuts are complete, and whether the sequence is correct.

③ Check whether the routing direction and outlet position meet requirements.

④ Measure the resistance value between positive and negative poles of DC circuit, and the resistance value between phases of AC circuit. Make sure there is no short circuit or open circuit.

⑤ Check whether the color of AC line is standard, and whether the safety signs are complete.

⑥ Check whether the connection stability, line sequence and polarity of DC output and battery are correct.

⑦ Check whether the electrical components are connected and fixed firmly.

⑧ Check the corresponding relationship between DLC optical cable ports and sectors, and make sure there is no error.

⑨ Check whether the grounding wire is connected correctly and firmly.

⑩ Whether the wiring is neat and whether the cable binding meets requirements.

(3) Inspection of frame installation.

① Check whether the horizontal, vertical and stable installation of the frame meet requirements.

② Check whether all bolts are tightened, whether plain washers, spring washers and nuts are complete, and whether the sequence is correct.

③ Check whether the routing direction and outlet position meet requirements.

④ Measure the resistance value between positive and negative poles of DC circuit, and the resistance value between phases of AC circuit. Make sure there is no short circuit or open circuit.

⑤ Check whether the color of AC line is standard, and whether the safety signs are complete.

⑥ Check whether the connection stability, line sequence and polarity of DC output and battery are correct.

⑦ Check whether the electrical components are connected and fixed firmly.

⑧ Check the corresponding relationship between DLC optical cable ports and sectors, and make sure there is no error.

⑨ Check whether the grounding wire is connected correctly and firmly.

⑩ Whether the wiring is neat and whether the cable binding meets requirements.

4. Attentions

It is necessary for students to strictly follow the safety precautions when using the equipment.

5. Conclusion of the task

_____

_____

_____

_____

Task 2   Installation of RRU System

※Related knowledge

The RRU system consists of RRU (including installation components), smart antenna, upper jumper, NB-RRU optical fiber, RRU power line, and RRU grounding kit, etc.

The exterior of Remote Radio Unit (RRU) is shown in Figure 2-2-1.

Figure 2-2-1　Exterior of TDRU338D

※ Practice

1. Preparation before the task

1) Inspection and verification of installation environment

To ensure the EMB5116 TD-LTE compact base station performs safely, stably and reliably, the machine room where the equipment is installed should be set in the place that can provide the base station with a good working environment, rather than in the place with high temperature, dust, noxious gas and inflammable and explosive materials; In addition, places with strong vibration and noise should be avoided; stay away from the substation and high-voltage transmission line; The building structure, heating and ventilation, power supply and water supply, lighting and fire protection of the machine room should be designed and constructed in accordance with relevant national and industrial standards.

When applied for outdoor macro coverage, the machine room should also guarantee the wall mounting space (height×width×depth) for DC power lightning protection box: 750mm×500mm×500mm.

The inlet of antenna and feed line of the machine room should be set on the top of the nearby wiring rack, and should be equipped with waterproof and sealing devices.

The doors and windows of the machine room should have good sealing and dust-proof functions and anti-theft devices.

2) Preparation of the tools

Tools required for installation are shown in Table 2-1-1.

2. What to do in this task

Install the RRU equipment.

3. Training process

There are two ways to install the Remote Radio Unit (RRU) chassis: clamping-pole mounted and wall-mounted.

1) Clamping-pole-mounted RRU chassis

The general installation fixture can be used on the pole with a diameter of 50mm-114mm. Refer to Figure 2-2-2 for the installation sequence of each component.

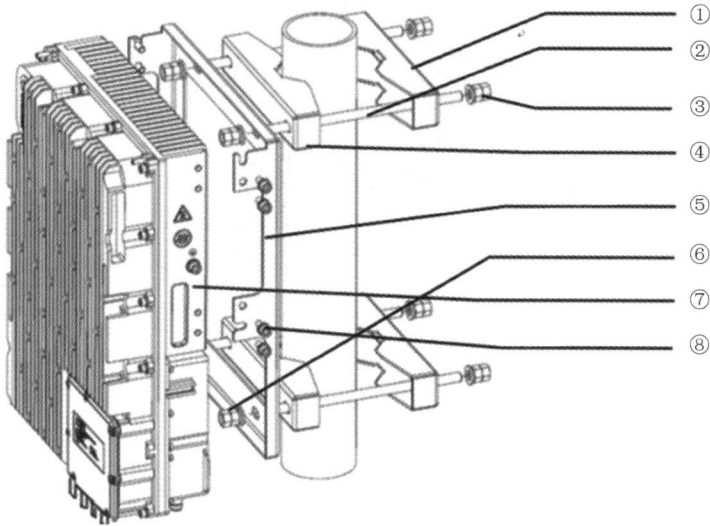

Figure 2-2-2   Diagram of Component Installation

The specific installation steps are as follows:

(1) Remove the mounting bracket: loosen a total of 8 sets of bolts connecting RRU and the mounting bracket, and remove the mounting bracket from the RRU, as shown in Figure 2-2-3.

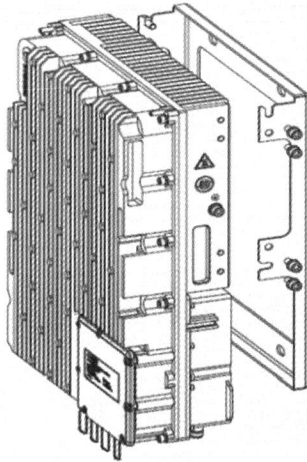

Figure 2-2-3   Diagram of Removing the Mounting Bracket

(2) Firstly, assemble the nuts (two), spring washer, plain washer and a clamp hoop successively on M12×240 screw in order to make one side of the fixture closed and the other side open.

(3) Pass the open side through clamping pole, another clamp hoop, the RRU mounting bracket, the plain washer, the spring washer, and the nuts (two), and then fix the fixture at the predetermined position of the pole, as shown in Figure 2-2-4; the upper planes of the two hoops in two sets are required to be at the same level, and the bolts on one side of the RRU mounting bracket are required to stretch out from the nut by 3-5 threads to avoid collision when RRU is installed.

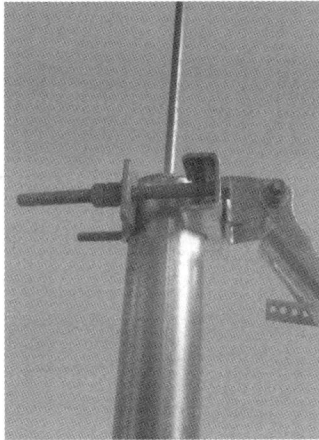

Figure 2-2-4　Diagram of Clamp Hoop Installation

(4) After adjusting the bolts and hoops according to the requirements listed in step ③, tighten the nuts to the corresponding torque, ranging from 10-12N·m, to fix the hoops; the sequence of tightening the nut is: first, tighten the inner nut of the two nuts at each end of the bolt according to the requirements for torque, and then tighten the outer nut. It is forbidden to tighten the two nuts at the same time.

(5) Hang the RRU chassis on the back frame and tighten eight sets of bolt components of the RRU chassis.

2) Wall-mounted RRU chassis

General installation fixture can be used for wall mounting as well. Refer to Figure 2-2-5 for the installation sequence and position of each component.

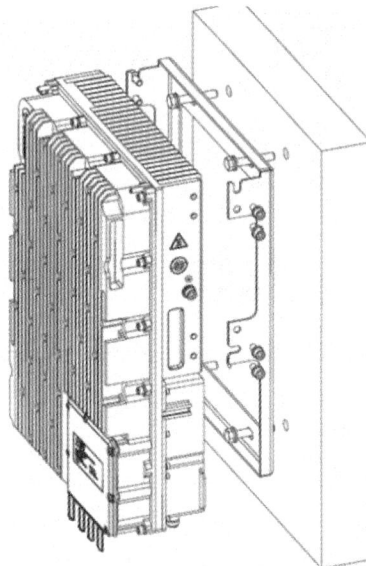

Figure 2-2-5　Diagram of Component Position of Wall-mounted RRU

Installation steps:

(1) Remove the mounting bracket: loosen 8 sets of bolts connecting the RRU and the

mounting bracket, and remove the mounting bracket from the RRU, as shown in Figure 2-2-3.

(2) Place the removed RRU mounting bracket at the predetermined installation position, as shown in (a) of Figure 2-2-6, keep the upper plane of the mounting bracket horizontal, mark the mounting holes of the anchor bolts with a marker, remove the mounting bracket, and drill the mounting holes of the anchor bolts.

(3) As shown in (b) of Figure 2-2-6, first place the anchor bolt sleeves into the installation holes on the wall, then fix the installation bracket on the wall with four sets of anchor bolts, and then screw the bolts half a circle after the spring washers are pressed flat.

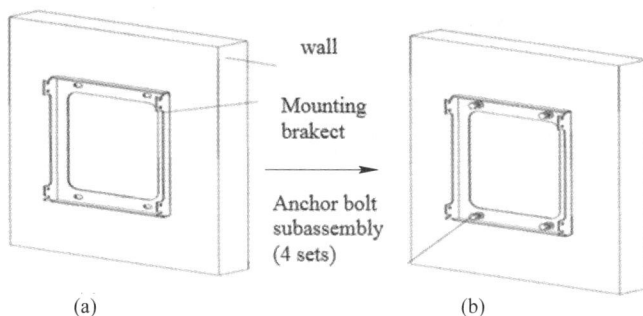

wall

Mounting brakect

Anchor bolt subassembly (4 sets)

(a)　(b)

Figure 2-2-6　Diagram of Fixing the Mounting Bracket

(4) Hang the RRU chassis to the installed bracket, and then tighten the eight screw subassemblies of the RRU chassis.

3) Antenna installation

The antennas that can match with TDRU338D contain 8-element dipole sector smart antenna and 8-element omni-directional smart antenna (ring-arrayed). The installation of 8-element dipole sector smart antenna is mainly introduced in the following part.

Installation steps:

(1) Unpack the antenna packaging carton, and lay the back panel of the antenna upward on a flat, open and clean ground to avoid scratching. Refer to Figure 2-2-7 for the overall composition and structure of the antenna.

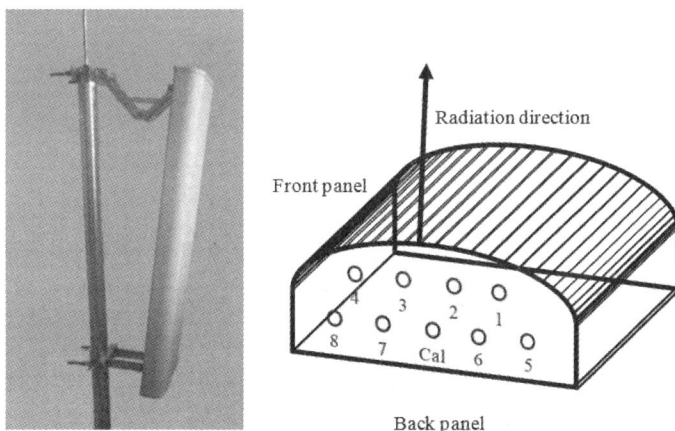

Radiation direction

Front panel

4
3　2　1
8　7　Cal　6　5

Back panel

Figure 2-2-7　Diagram of 8-element Smart Antenna

(2) Take out the installation structural parts, assemble and install them to the position of the antenna backplane according to the instructions, with one set for the upper part and lower part respectively, as shown in Figure 2-2-8.

Figure 2-2-8   Antenna Installation Structural Parts

(3) Put 9 upper jumpers through the outdoor waterproof heat-shrink tubings, and connect them to 1, 2 , ..., 8 and CAL port on the lower end of antenna, and then tighten the connector nuts. Figure 2-2-9 shows the lower end of the antenna.

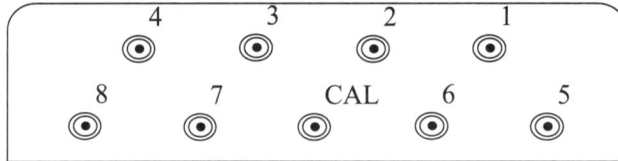

Figure 2-2-9   Diagram of the Lower End of the Antenna

(4) The standing wave ratio of each port should be tested to ensure that it is no greater than 1.4.

📖 Notice:

If the standing wave ratio of a port is greater than 1.4, then disconnect the upper jumper of the corresponding port, and test the standing wave ratio of the antenna port and the upper jumper respectively.

(5) Install the antenna on the vertical antenna clamping pole, adjust the antenna to the designed height, and then initially tighten the clamp nuts to prevent the antenna from sliding and falling, as shown in Figure 2-2-10.

Figure 2-2-10   Antenna Installation Complete

⚠Caution:

Antenna installation is generally performed high above the ground. When fixing the antenna, the installation personnel should wear protective equipment and be properly fixed to protect personal safety.

(6) According to the values of design drawings, compass and goniometer are used to adjust the horizontal azimuth and pitch angle of the antenna. After the adjustment, lock all fixing nuts, as shown in Figure 2-2-11.

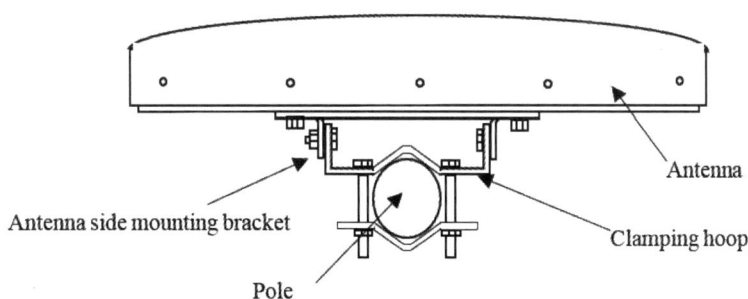

Figure 2-2-11   The Top View of the Antenna After Installation

📖 Notice:

Omni-directional smart antenna needs to be installed vertically, and others are installed the same way as sector smart antenna.

⚠ Caution:

The antenna is the source of electromagnetic signal, so the radiating surface of the antenna should not be blocked within a certain range of sight distance.

4) Standard for installation and arrangement of wires

(1) Standard for upper jumper installation and arrangement. EMB5116 TD-LTE adopts a 400 type upper jumper with N-type male connectors at both ends. The minimum bending radius is 25.4mm and the minimum alternating bending radius is 101.6mm.

Installation steps:

① Connect one end of the upper jumper to the N-type female connector at the lower end of the antenna, and connect the other end to the N-type female connector at the lower side of TDRU338D. ANT1, ANT2, …, ANT8 and CAL port of the chassis should correspond to 1, 2, …, 8 and CAL port of the antenna, as shown in Figure 2-2-12.

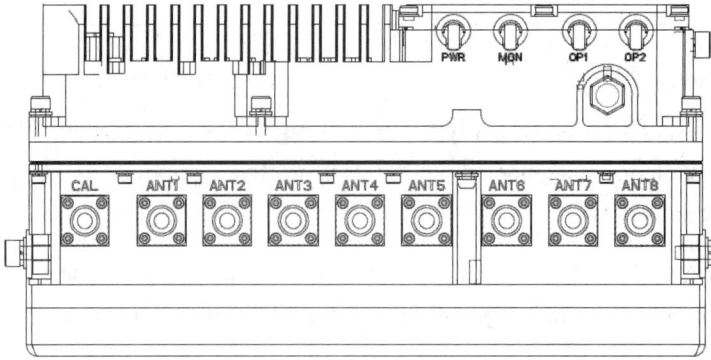

Figure 2-2-12　View of the Lower End Face of TDRU338D

📖 Notice:

Based on the length of the upper jumper, feed line clips or black nylon binding tape should be used to fix it, with a feed line clip set at every 400mm. Where feed line clips cannot be used, the upper jumper should be bound with black nylon binding tape of outdoor type, with an interval of 400mm.

② Heat-shrink tubings are used to make the joint between the upper jumper and the antenna and TDRU338D waterproof.

📖 Notice:

The waterproof heat-shrink tubing is a black 28/9 double-walled polyolefin heat-shrink tubing with adhesive, which can be heated by an electric heat gun or a fuel spray gun. It is required to heat from the end near the antenna or TDRU338D until a small amount of glue overflows at both ends. Make sure that tubing is sealed and shrunk completely. Heat-shrink tubings should be shrunk evenly, as shown in Figure 2-2-13.

The effect of a heat-shrink tubing after its shrinking is shown in Figure 2-2-13. In the process of heating, it is necessary to always push the heat-shrink tubing to one side of the top surface of RRU.

Figure 2-2-13　Installation Diagram of a Heat-shrink Tubing

③ Bind the aluminum signboard. Bind the corresponding aluminum signboards to both ends of the upper jumper with a 2.5mm×100mm black binding tape.

📖 Notice:

The binding requirements of the signboard are as follows: the signboard should be 200 mm away from both ends of the upper jumper, with the font direction facing upward. Two binding tapes are used for each signboard, and the excess length of the tape should be cut off, leaving another 3-5 buckles to spare.

⚠Caution:

It is strictly prohibited to fold the upper jumper into a right angle when laying it. Rather, bending or turning should be avoided as much as possible. When it is really necessary to turn, the minimum bending radius of the jumper should be met, and then the convenience and beauty of wiring should be considered.

If there is excess length of a jumper, the excess part should be coiled and placed on the wiring rack near RRU and be bound with nylon tape.

Jumpers should not be intertwined and crossed, and the routing should be straight and neat.

The marks at both ends of the jumper should be clear, neat and easy to inspect. The contents on the tags should match with the antenna and RRU ports.

Foam cushion or other protective materials must be provided when the jumper turns a corner, to prevent the jumper from scratching due to the edge of the wiring rack.

(2) Standard for installation and arrangement of NB-RRU optical fiber. NB-RRU optical fiber is a two-cored optical fiber, and the installation position is shown in Figure 2-2-14.

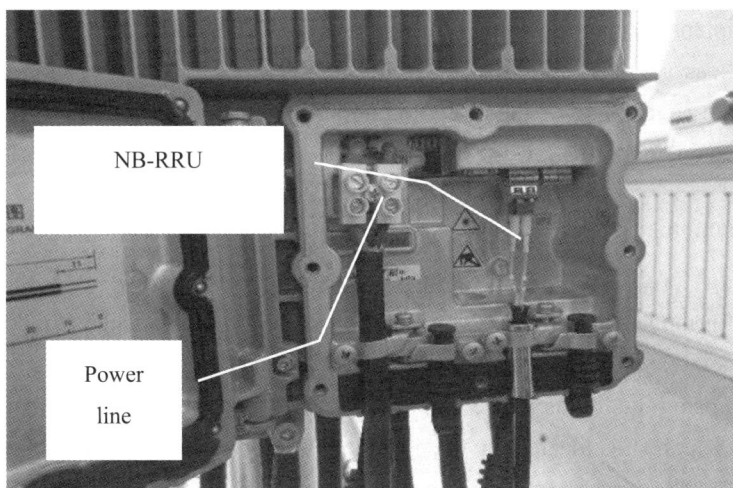

Figure 2-2-14   Installation Position of the Fiber

The optical cable assembly provides the data connection path between EMB5116 TD-LTE and TDRU338D, and its profile is shown in Figure 2-2-15. Through the DLC plug of crimping end (A end) and branching end (B end), the duplex LC photoelectric module of eNode B and RRU equipment is interconnected.

Figure 2-2-15　Optical Cable Assembly

Installation steps:

① Open the cover plate of TDRU338D operation and maintenance window. Before opening the maintenance window, make sure that there is no large amount of visible floating dust, and water vapor within 5m that may enter the maintenance window, as shown in Figure 2-2-14.

② Open the crimping half ring on the corresponding installation position of the optical fiber in the maintenance window, as shown in Figure 2-2-14.

③ Take out the optical cable assembly, crimp the T-shaped metal sheath at the A end of the optical cable with the crimping half ring, put it in the position of OP1 (left side) by default, and fasten the screws on the crimping ring.

④ Remove the DLC dust cap at the end of the optical fiber, (do not remove the dust cap if the optical fiber is not to be installed), and carefully plug the DLC into the OP1 optical module of the RRU maintenance window. Plug corresponding LC connector into optical module Rx and Tx, and avoid excessive bending of the pigtail during the process.

⑤ Close the cover plate of the operation and maintenance window and tighten all screws on the maintenance window.

⚠ Caution:

The tightening torque of all screws at the maintenance window is 1.4N·m, and the requirement for the installation site is "all the screws at the maintenance window should be screwed half a circle after the spring washers are pressed flat".

Before closing the maintenance window, check that the sealing rubber ring around the maintenance window is located in the installation groove on the maintenance window. After tightening the screws on the maintenance window, check that the sealing ring does not squeeze through the maintenance window.

Close the maintenance window immediately after installing the fiber.

⑥ Fix the optical cable on the wiring rack with the fixing clips and the reinforced sheath of the optical fiber.

📖 Notice:

With the fixed interval set at 800mm, the reinforced sheath of the fiber can ensure that the cable is firmly fixed on the fixture, and at the same time, cables should avoid being squeezed and damaged by the fixture. If the vertical part of the optical fiber is too long, the fixed interval should be shortened. The fixture interval can be adjusted according to the cross arm interval of the wiring rack. Sufficient margin should be left for optical fiber at RRU end.

---

⑦ The optical fiber passes through the feed line window, and the DLC plug at the B end should be plugged into the corresponding optical fiber port on the BPOG board of the EMB5116 TD-LTE equipment.

⚠ Caution:

The bending radius of the optical cable should be 20 times larger than the diameter of the optical cable. It is not allowed to fold the optical fiber into a right angle. Optical fiber pairs should be neatened and bound. It is strictly prohibited to lay, bend, and bind optical cables in the same way as the electric cables.

When the plug or the jumper of a socket is not to be used, do not open the dust cap of the optical fiber connector; if it must be opened, cover the dust cap of the plug or socket as soon as possible when the operation is over, so as to avoid dust entering the optical fiber connector, polluting the end face of the optical fiber pin, and thus resulting in the decline or damage of the product performance.

The wiring arrangement for the optical cables should be natural, knots not allowed. It is prohibited to violently pull and drag optical cables when they are intertwined. They should be bound mildly. The optical fiber fixed on the cross bar of the wiring rack should be bound once every 400mm. Turns should be reduced. Given the convenience and beauty of the routing, the optical fibers between cabinets can be laid through the wiring rack or directly on the top of the machine.

It is strictly prohibited to drag the optical cables by holding the connector and corrugated pipe (used to protect the connector).

It is strictly prohibited to violently drag the optical cables if they are stuck during the construction.

It is prohibited to lay or drag optical cables on the surfaces of sharp objects.

When optical cables pass through the pipeline, it is necessary to use the traction rod. Seal the LC head of optical cable with its original plastic bag, and fix it with wide waterproof tape along the traction rod. The fixed length should be greater than 0.5m, with knots prohibited, and be wrapped with three layers of waterproof tape.

In the construction process, special treatment should be carried out if the optical fiber connector (LC) is immersed in water. The treatment should be as follows: wash the connector with clean water first, and after the connector dries completely, wipe the end face of the pin with clean degreasing cotton swab dipped in anhydrous alcohol. After the alcohol volatilizes, the optical connection can be carried out.

During the process of laying the optical fiber, if the length is redundant, the redundant part should be placed on the wiring rack to be bound. Do not intertwine the optical fibers, otherwise it is not convenient to identify.

The marks at both ends of the optical fiber should be clear, and the contents on the tag should describe the corresponding relationship between RRU and optical fiber port of the BPOF board.

When the optical fiber is laid around the corner, foam pad must be provided, to prevent the optical fiber from scratches from edges and corners.

(3) Installation of four-cored optical fiber: refer to Figure 2-2-16 for the steps of installing the four-cored optical fiber.

① The DLC plug marked with Rx1 and Tx1 should be plugged into the optical module marked with OP1 on RRU (optical module on the left side), and the DLC plug marked with Rx2 and Tx2 should be plugged into the empty optical module installation hole or optical module marked with OP2 on RRU (generally, the position of OP2 is not installed with optical module).

② The DLC plug marked with Rx1 and Tx1 on the RRU side correspond to the DLC plug marked Tx0 and Rx0 on the BBU side, which means that they are the same pair of optical fibers; The DLC plug marked Rx2 and Tx2 on the RRU side correspond to the DLC plug marked Tx1 and Rx1 on the BBU side, which means that they are the same pair of optical fibers.

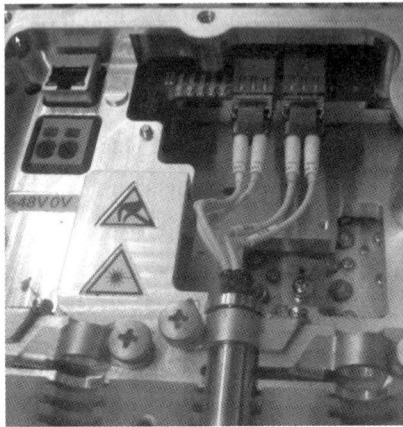

Figure 2-2-16   Installation Diagram of the Four-cored Optical Fiber

(4) Standard for installation and arrangement of RRU power line.

Installation steps are as follows:

① Cut the power line to its required length, as shown in Figure 2-2-17.

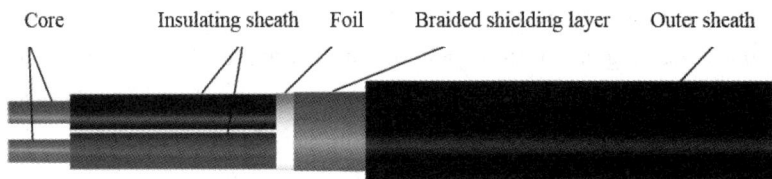

Core   Insulating sheath   Foil   Braided shielding layer   Outer sheath

Figure 2-2-17   The Diagram of Two-cored 6mm2 Power Line

② If the power line end at RRU side is made on site: the power line should be stripped off accordingly—the outer sheath should be stripped off by 65mm; the shielding layer—the braided shielding layer should be stripped off by 15mm. Then, the shielding layer should be folded over to the outer sheath of the power line. If there is aluminum foil or plastic tape, it must be cut off. The insulating sheath (red and blue) should be stripped off by 8mm. The copper wire of the core should be neat, and there should be no broken copper conductor or braided shielding layer copper wire left on the power line.

③ Install the heat-shrink tubings of the power line terminal on the red and blue insulation sheath of the power line respectively.

④ Crimp the right angle terminal at the end of the power line with the six-square-jaw crimping pliers. The direction of the two terminals should be the same when the power line is in the natural state, and the crimping should be firm and reliable.

⑤ Push the heat-shrink tubing to the corner of the terminal, and use the heat gun or other heat source to shrink the heat-shrink tubing in place, as shown in Figure 2-2-18.

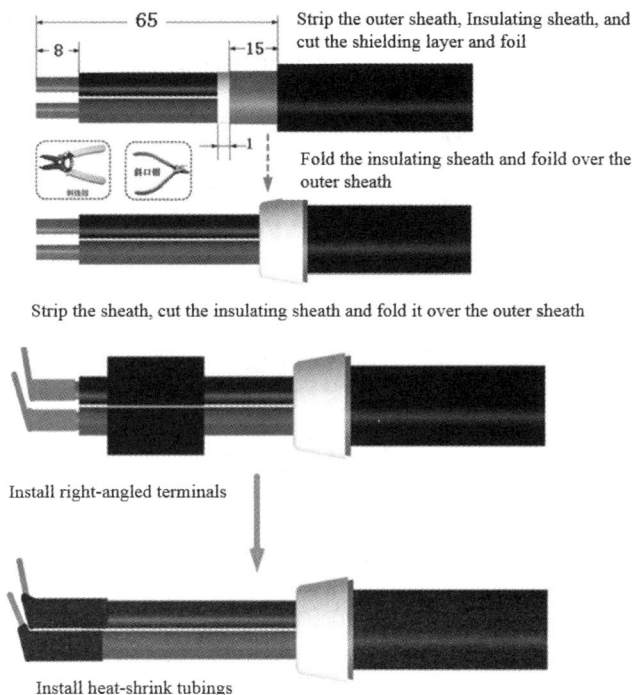

— 65 —   Strip the outer sheath, Insulating sheath, and cut the shielding layer and foil
8   15
1   Fold the insulating sheath and foild over the outer sheath

Strip the sheath, cut the insulating sheath and fold it over the outer sheath

Install right-angled terminals

Install heat-shrink tubings

Figure 2-2-18   Diagram of Making Two-cored 6mm2 Power Line

⑥ Open the maintenance window of RRU, loosen the fastening screws on the semicircle crimping ring corresponding to the installation position of the power line, install the folded part of the power line shielding layer into the semicircle groove, then compress the power line (including the shielding layer) with the semicircle crimping ring, and finally fasten the screws; the front part of the outer sheath of the power line should be at least 2mm longer than the semicircle crimping ring, as shown in Figure 2-2-16.

⑦ Plug the connector of the power line into the hole of the terminal in the equipment. No tools are needed during the plugging process. The red (outer sheath) wire should be plugged into the hole marked 0V, and the blue (outer sheath) wire should be plugged into the hole marked −48V. Make sure that they are installed in place and that no single copper wire is scattered out of the connection hole.

⑧ Check after the installation to ensure that no broken copper wire falls into the RRU maintenance window.

⑨ Close the cover plate of the operation and maintenance window and tighten all the screws on the maintenance window.

If the optical fiber and power line are installed at the same time, install power line first and then the optical fiber.

(5) Standard for installation and arrangement of RRU grounding wire.

Installation steps:

① The 16mm^2 RVVZ yellow green power line is used when TDRU338D chassis is grounded, and the appropriate length of the power line can be cut according to the grounding location on the project site.

② Crimp the 16mm^2 cable lugs with M8 bolt hole at both ends, and the crimping part between the cable lug and the cable should be wrapped with narrow insulating tape for protection, as shown in Figure 2-2-19.

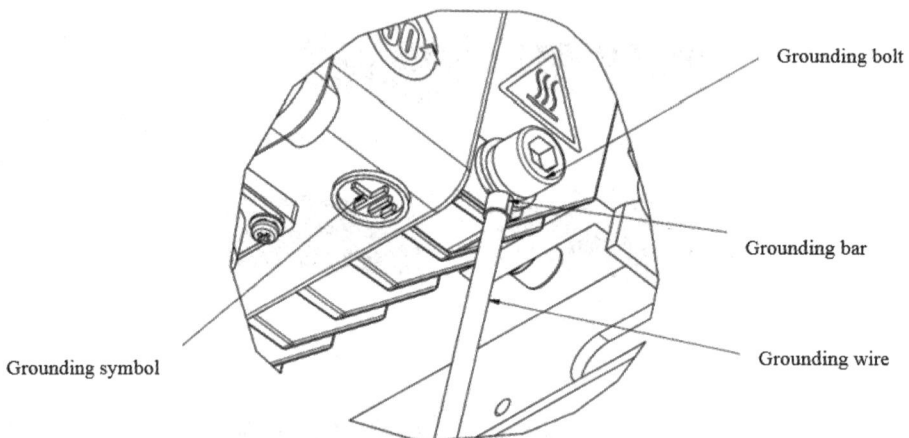

Grounding bolt

Grounding bar

Grounding wire

Grounding symbol

Figure 2-2-19　Diagram of Chassis's Grounding

③ One end of the grounding cable is reliably connected to the chassis's grounding point with hexagon socket head cap screw mounted on the RRU chassis, and the other end is connected to the grounding bus bar or grounding point on the project site.

(6) Binding and fixing cables. The external cables of TDRU338D mainly contain upper jumper, power line, RRU optical fiber and grounding cable.

① The power line is RVVP $2 \times 6 mm^2$ shielding power line, with an outer diameter of (11.2 ±0.1)mm. It can be fixed on the wiring rack with a standard 1/4 feed line clip, as shown in Figure 2-2-20.

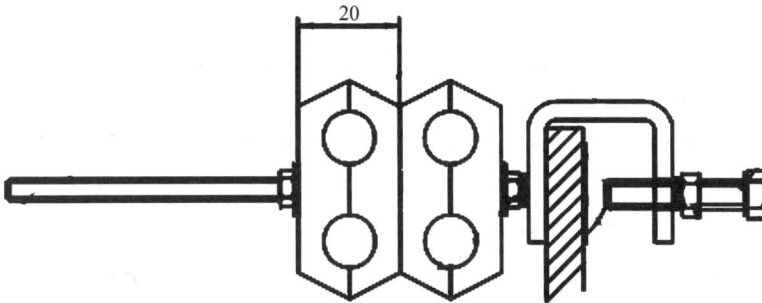

Figure 2-2-20   Diagram of Feed Line Clip's Structure

② The outer diameter of the optical cable is 7mm. The diameter of the cable is increased to 11mm by using the reinforced sheath at the fixed position where the same type of the feed line clip is set. Refer to Figure 2-2-21 for the dimension of the reinforced sheath.

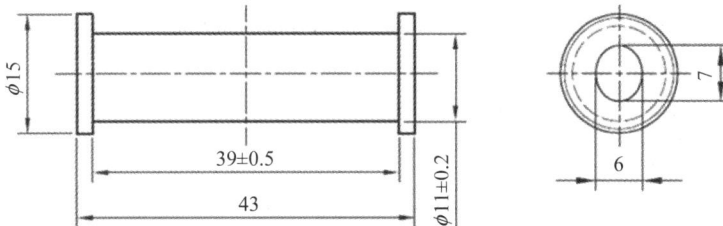

Figure 2-2-21   Dimension and profile of the Reinforced Sheath

③ The reinforced sheath is a natural rubber cut along the axial direction. Refer to Figure 2-2-22 for the usage.

Figure 2-2-22   Diagram of the Usage of the Reinforced Sheath

④ The upper jumper is made of type 400 RF flexible cable, which can be fixed directly with feed line clips.

⑤ For places where feed line clips cannot be used, 7.7×370 black nylon tape can be used outdoors, while indoors 5×250 white tape can be used. The buckles should face the same direction when the feed line is bound with the tape. The excess length of the indoor tape should be cut off along the buckle, while the excess length of the outdoor tape should be cut off, leaving another 3-5 buckles to spare. The interval can be adjusted properly according to the cross bar of the wiring rack on the site. Generally, the interval for binding ordinary cables should be no greater than 400mm.

5) Installation inspection

Installation inspection should at least include the following steps:

(1) Check whether the installation of RRU and antenna are firm and reliable, and whether the installation position meets the requirements.

(2) Check whether the type and quantity of all devices of TDRU338D subsystem are correct.

(3) Check whether all the fasteners (like screws, nuts, washers, etc.) of TDRU338D subsystem, ranging from the antenna to the equipment are complete and fastened as required.

(4) Check whether the waterproof treatment of all connectors and grounding casings meet the requirements and are reliable.

(5) Check whether the wiring is neat, whether the cable binding meets the requirements, and whether the tag binding meets the requirements.

(6) Check whether the installation direction of all connectors is correct.

(7) Whether the sheath of all cables is damaged, and whether the minimum bending radius of cables meets the requirements.

(8) Clean up the waste and count tools.

6) Installation of other RRUs

The steps and methods for installing other models of RRU are the same as those for TDRU338D.

4. Attentions

It is necessary for students to strictly follow the safety precautions when using the equipment.

5. Conclusion of the task

_____

_____

_____

_____

_____

_____

Task 3   Installation of GPS

※ Related knowledge

GPS consists of its antenna (including its installation components), feed line, lightning arrester, jumper, down jumper, SCTE board (in BBU), feed line grounding kit, lightning arrester grounding kit, etc. Figure 2-3-1 describes a standard structure.

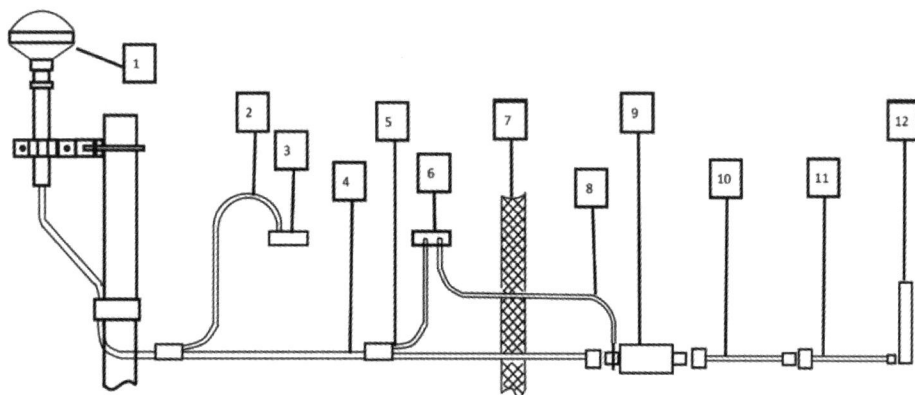

1—Antenna (installation components included); 2—Feed line grounding kit; 3—Grounding bar;

4—Feed line; 5—Feed line grounding kit; 6—Grounding bar; 7—Feed line window;

8—Lightning arrester grounding kit; 9—Lightning arrester; 10—Jumper; 11—Downjumper; 12—SCTE board

Figure 2-3-1   Diagram of GPS System's Structure

※ Practice

1. Preparation of the tools

Tools required for installation are shown in Table 2-1-1.

2. What to do in this task

Install GPS.

3. Training process

1) Installation of the antenna

(1) Clamping-pole-mounted antenna.

Installation steps are as follows:

① Fix the installation pipe with the fastener. The installation pipe should be vertical and firm. After the spring washer is pressed flat, it should be fastened for half a circle.

② Install the fastener, clamp and hoop according to the position and direction shown in

Figure 2-3-2. The hoop should pass through the clamp, fastener, plain washer, spring washer and nut in sequence. After the spring washer is pressed flat, it should be fastened for half a circle.

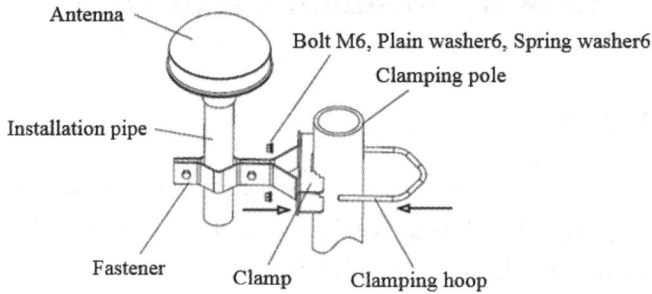

Figure 2-3-2   Diagram of Clamping-pole-mounted GPS Anttena

📖 Notice:

(1) The antenna should be higher than the top end face of the clamping pole.

(2) There is no requirement for the fastener's position on the pipe, nor the requirement for the clamping hoop's position on the clamping pole.

(3) Make sure that the installation is firm and reliable and meets the requirement listed in (1).

(4) The GPS antenna must be kept vertical during installation (when the installation is carried out in the northern hemisphere, the antenna can be tilted 2°~3° to the south).

(5) The GPS antenna must be within the lightning protection range of the lightning rod, which is a cone of 45°.

(6) Keep the GPS antenna away from elevators, air conditioning, electronic equipment or other electrical appliances during installation. The antenna should be kept at least 2m away from metal objects.

(7) The horizontal distance between GPS antenna and other transmitting antenna (back) should be greater than 5m. When it is installed with the base station antenna vertically, the vertical distance between the bottom of antennas should be greater than 3m.

(8) It is forbidden to install GPS antenna in the main lobe of radiation antenna of base station and other systems.

(2) Wall-mounted antenna.

Installation steps are as follows:

① Fix the installation pipe with the fastener. The pipe should be vertically and firmly installed. After the spring washer is pressed flat, it should be fastened for half a circle.

② Place the finished fastener, as shown in Figure 2-3-3, on a proper position on the wall. The proper position indicates that the antenna is higher than the upper end face of the wall, and remains vertical. At the same time, the installation position should meet the requirements of (3), (4), (5), (6) and (7) listed in the notice for clamping pole installation.

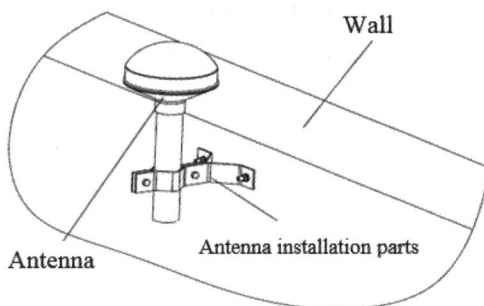

Figure 2-3-3   Diagram of Wall-mounted GPS

③ Based on the position introduced in ②, determine and drill the installation holes for anchor bolts through the holes on the fastener, and then place the anchor bolts.

④ From one side of the wall, install in sequence the fastener, plain washers, spring washers and nuts, and then fasten the anchor bolts.

2) Installation of the non-original clamping pole

The installation of the non-original clamping pole is divided into two steps: the non-original clamping pole installation and GPS antenna installation. The installation of the non-original clamping pole can be further divided into floor mounting, wall mounting, tower mounting and other typical scenarios.

(1) Steps for installing the floor-mounted non-original clamping pole are as follows:

① According to step (a) shown in Figure 2-3-4, connect the base of GPS mounting pole and L-shaped bracket with fasteners (four sets). The bottom surfaces of the two parts must be in the same plane. After the spring washers are pressed flat, they should be further tightened for half a circle.

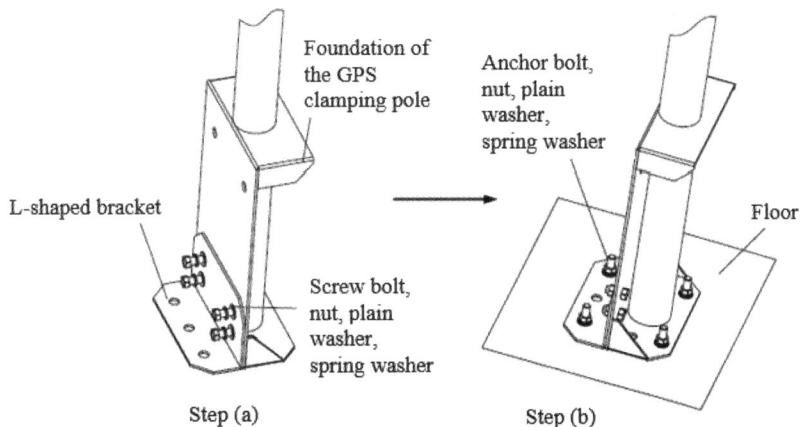

Figure 2-3-4   Diagram Floor-mounted Clamping Pole

② Select the level ground, drill the installation holes for anchor bolts on the ground according to the position of step (b) in Figure 2-3-4, and fasten the anchor bolts (four sets in total) in a diagonal sequence to ensure the connection is firm and reliable.

(2) Steps for installing the wall-mounted non-original clamping pole are as follows:

① Select the vertical face of a straight wall, and drill installation holes for anchor bolts on the wall according to Figure 2-3-5.

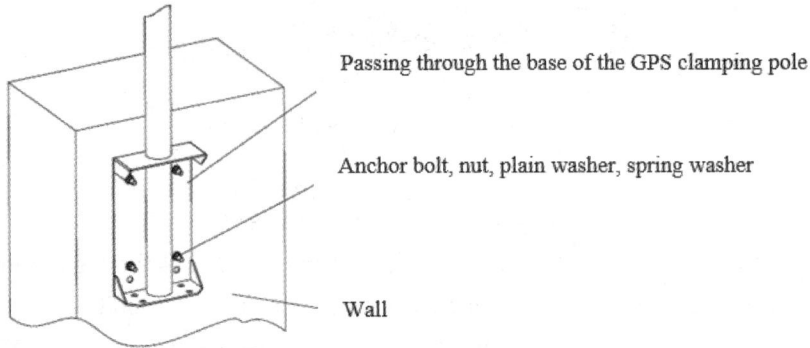

Passing through the base of the GPS clamping pole

Anchor bolt, nut, plain washer, spring washer

Wall

Figure 2-3-5　Diagram of Wall-mounted Clamping Pole

② Install the anchor bolts, and pass them through the base holes of GPS pole, plain washers, spring washers and nuts in sequence, and then fasten the anchor bolts in diagonal sequence.

📖 Notice:

After the installation, it is necessary to ensure that the GPS clamping pole is vertical.

(3) Steps for installing the tower-mounted non-original clamping pole are as follows:

① Select the straight and horizontal part of the steel structure of the iron tower, and drill two installation holes according to the position shown in Figure 2-3-6. Horizontally, the center distance between the installation holes is 72mm, and the installation holes can be any two holes at the same horizontal line on the base of GPS installation pole (including on the side and the bottom), based on the steel structure.

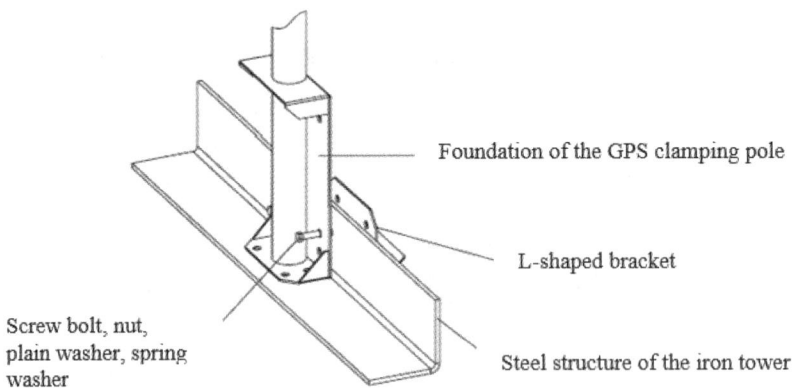

Foundation of the GPS clamping pole

L-shaped bracket

Screw bolt, nut, plain washer, spring washer

Steel structure of the iron tower

Figure 2-3-6　Diagram of Tower-mounted Clamping Pole

② Install the bolts, and pass them through the holes on the base of GPS installation pole, steel structure, L-shaped bracket, plain washers, spring washers and nuts in sequence. After the spring washers are pressed flat, they should be tightened for half a circle.

📖 Notice:

After the installation, it is necessary to ensure that the GPS clamping pole is vertical.

3) Installation of cables and devices

(1) Installation of type N connector on a feed line, as shown in Figure 2-3-7. This process is applicable to the installation of type N connector (male and female) on 400DB feed line and 600DB feed line.

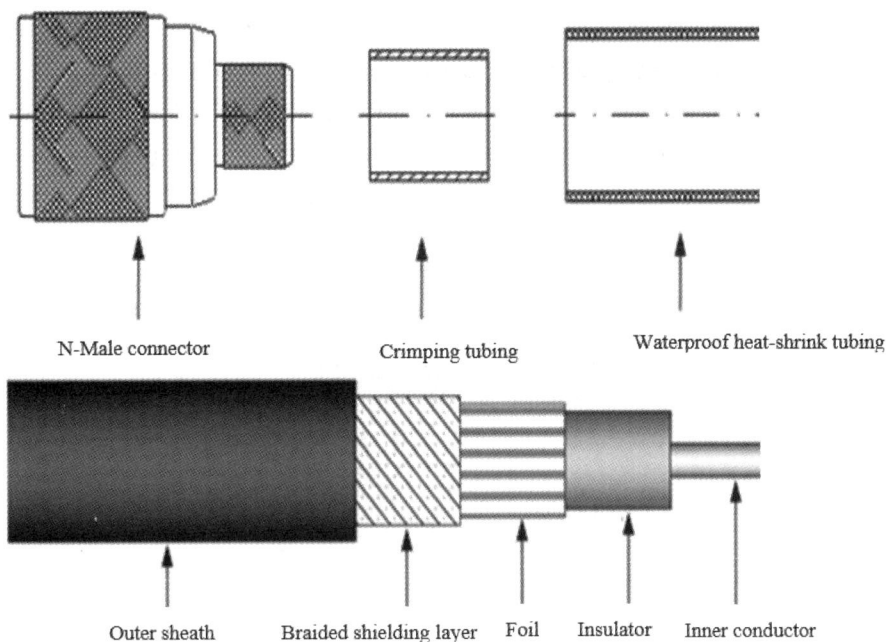

N-Male connector   Crimping tubing   Waterproof heat-shrink tubing

Outer sheath   Braided shielding layer   Foil   Insulator   Inner conductor

Figure 2-3-7   Structural Diagram of Type N Connector and GPS Feed Line

Installation steps are as follows:

① Strip the sheath of GPS feed line, partly exposing the inner conductor, and the reserved length of weave layer and inner conductor should be 14mm and 6mm respectively, and then use file or angle gauge to process the inner conductor. Refer to Figure 2-3-8 for length and radian angle.

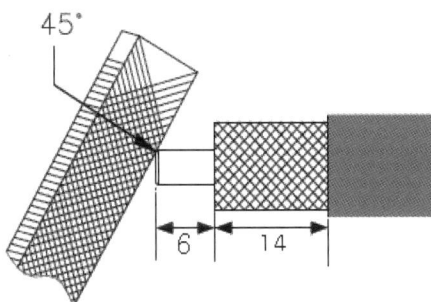

45°

6   14

Figure 2-3-8   Length of the Stripped Wire and the Radian Angle

② Pass the feed line through the waterproof heat-shrink tubing and crimping tubing in sequence, and push the feed line into the type N connector, as shown in Figure 2-3-9.

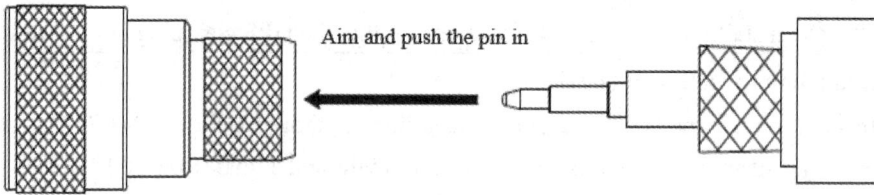

Aim and push the pin in

Figure 2-3-9　Diagram of Pushing in the Feed Line

③ Install the feed line and connector in position, as shown in Figure 2-3-10. Use special tools to crimp firmly the crimping tubing. The hexagonal crimping pliers with the opposite sides measuring 10.897mm (0.429") should be used to crimp if the type 400 feed line is involved, and the hexagonal crimping pliers with the opposite sides measuring 15.5mm (0.61") should be used to crimp if the type 600 feed line is involved. Refer to Figure 2-3-11 for the appearance after crimping.

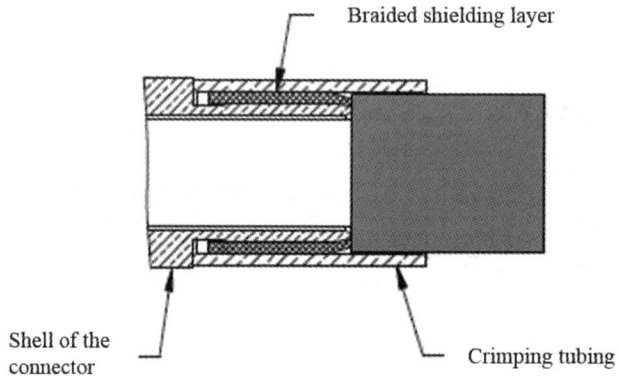

Braided shielding layer

Shell of the connector

Crimping tubing

Figure 2-3-10　Feed Line Installed in Position

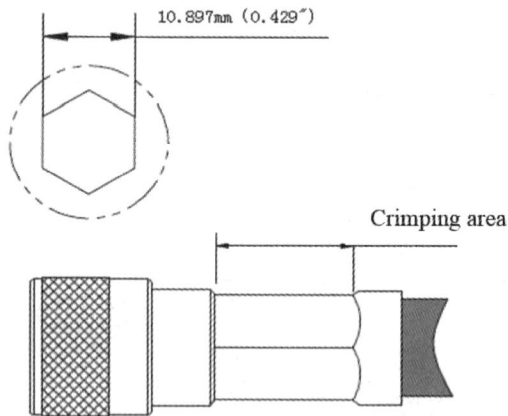

10.897mm (0.429″)

Crimping area

Figure 2-3-11　Diagram of Crimping Tubing after Being Crimped

④ Push the heat-shrink tubing in position , and then use the heat gun to make the tubing fully shrunk, with a small amount of glue squeezing at both ends, as shown in Figure 2-3-12.

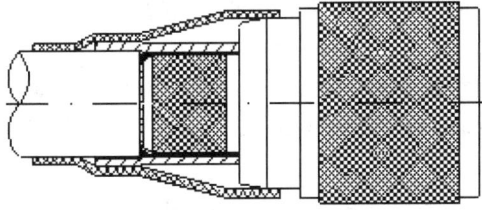

Figure 2-3-12　Diagram of Installing Heat-shrink Tubing

(2) Connection of feed line and antenna.

① Pass the feed line through the antenna installation pipe, and connect the type N connector (male) of the feed line to the port at the lower end of the antenna. Keep the feed line still, and turn the antenna to fasten the connection threads, as shown in Figure 2-3-13.

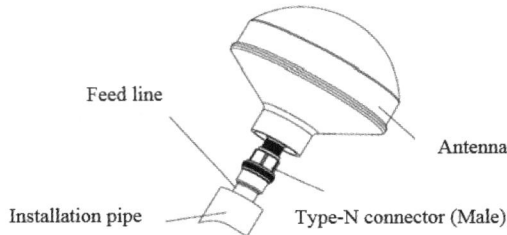

Figure 2-3-13　Diagram of Feed Line and Antenna Connection

② Tighten the installation pipe to the antenna, keep the antenna still, and then turn the antenna installation pipe to fasten the connection threads, as shown in Figure 2-3-14.

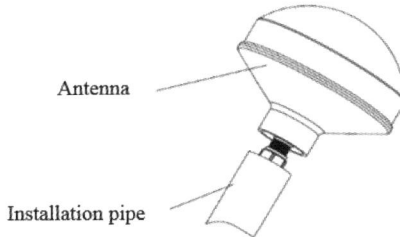

Figure 2-3-14　Diagram of Installation Pipe and Antenna Connection

(3) Installation steps of feed line grounding kit are as follows:

① Referring to the length of clamping hoop of the component 1 in Figure 2-3-15, strip the corresponding length of the feed line sheath without damaging the shielding layer.

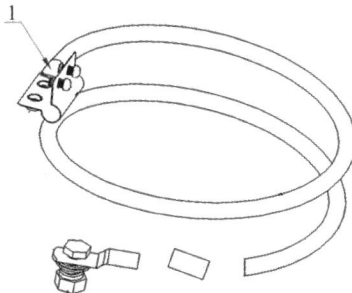

Figure 2-3-15　Diagram of Feed Line Grounding Kit

② Put the clamping hoop on the shielding layer of the feed line whose outer sheath is stripped off.

③ Fix the clamping hoop with machine screws. Referring to the sequence and direction shown in Figure 2-3-15, pass the screws through the clamping hoop, feed line and nuts in sequence, and then fasten the screws.

④ Waterproof treatment should be provided for the clamping hoop of feed line grounding by using adhesive tape.

⑤ Keep a proper length of the grounding cable according to the location of grounding point.

⑥ Crimp the copper lug ($10mm^2$-M8 single hole) at the other end of the grounding cable.

⑦ The crimped copper lug should be fixed on the grounding bar or tower body with bolts.

📖 Notice:

● When the length of the feed line is between 10m-60m, two grounding points need to be made. They are: the straight and horizontal part of the feed line 1m below the antenna, and the straight and horizontal part 1m before the feed line entering the feeder window.

● If the length of the feed line is less than 10m, the feed line grounding kit at the lower end of the antenna is not needed. Rather, the part in front of the feed line window needs to be grounded only.

● If the length of the feed line is greater than 60m, a grounding point should be added to the middle part of the feed line, which means that totally there will be three grounding points for the feed line.

(4) Installation of lightning arrester and its grounding kit. The lightning arrester and grounding kit should be installed at the nearest position after the feed line enters the feed line window. Installation steps are as follows:

① Remove the flange nuts and serrated washers on the lightning arrester, install the right-angled cable lugs of the grounding kit on the arrester, as shown in Figure 2-3-16. Then install the serrated washers and flange nuts in sequence, and tighten the nuts to fasten the cable lugs.

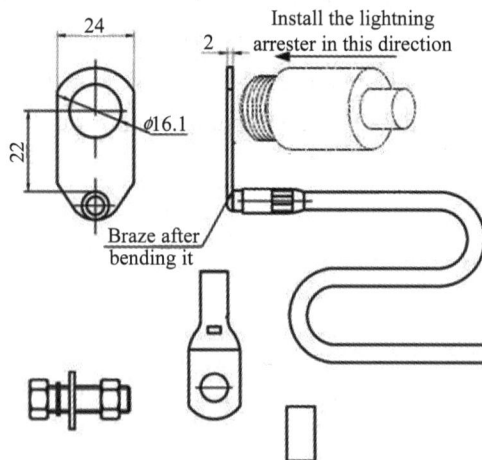

Figure 2-3-16   Diagram of the Lightning Arrester and Its Grounding Kit

② Connect the type N female connector at one end of the copper lug of the arrester with the type N male connector of the feed line, and then connect type N female connector at the other end of the arrester with the type N male connector of jumper or down jumper or the signal amplifier connection jumper, making sure that the connections are firm and reliable.

③ The other end of the lightning arrester grounding kit is connected to the indoor or outdoor grounding bus bar, and it is recommended that it be connected to the outdoor grounding bus bar by using $16mm^2$-M8 single hole copper lug.

(5) Feed line entering the window. Before the feed line enters the feed line window, make an elbow pipe to avoid water, and then tighten the fasteners and seal the feed line window, as shown in Figure 2-3-17.

Figure 2-3-17　Installation of Feed Line Window

⚠Caution:

It is necessary to make an elbow pipe to avoid rainwater seeping into the indoor main equipment along the feed line.

⚠Caution:

After laying the feed line, test the standing wave ratio and insertion loss of feed line channel separately. Standing wave ratio should not be greater than 1.3; insertion loss should be less than 20dB.

4. Attentions

It is necessary for students to strictly follow the safety precautions when using the equipment.

5. Conclusion of the task

Training Module 3   Installation of Auxiliary Equipment in Base Station and Optical Fiber Fusion Splicing

[Brief description]

In the field of wireless communication, the communication electronic equipment features high integration, vulnerability to over-voltage and over-current resistance, and sensitivity to external electromagnetic interference, especially to the interference of lightning electromagnetic pulse. Therefore, it is vital to install and use the lightning protection equipment. This training module focuses on the installation of lightning protection equipment and its grounding kit.

On the other hand, the communication between TD–LTE base station and other base stations, as well as the core networks relies on optical fibers. Optical fiber transmission enjoys the advantages of wide transmission frequency bandwidth, large transmission capacity, low loss, immunity to electromagnetic interference, small diameter of optical cable, lighter weight, and abundant raw material sources, etc. As the light is transmitted in the optical fiber, loss will be generated. Loss is composed mainly of the transmission loss of the optical fiber itself and the fusion loss of the optical fiber connector.

Once the optical cable is ordered, the transmission loss of the optical fiber itself is basically determined, and the fusion loss at the optical fiber connector is related to the optical fiber itself and on-site operations.

Reducing the fusion loss at the optical fiber connector can increase the transmission distance of the optical fiber repeater and amplifier, and decrease the attenuation of the optical fiber link. Therefore, it is very important to improve the fusion splicing quality of optical fiber and reduce the fusion loss.

Through the training of this module, trainees can master the installation of lightning protection facilities, grounding kits and optical fiber fusion splicing skills, so as to meet the basic requirements for installing and adjusting the mobile network.

[Elements of the training]

1. Knowledge objectives

(1) Master the classification of optical fiber connecting technology.

(2) Understand the working mechanism of optical fiber connecting.

(3) Understand the classification and principle of fusion splicing.

2. Ability objectives

(1) Master how to use fusion splicing tools.

(2) Master optical fiber fusion splicing as well as how to use the optical fiber terminal box.

3. Operational standards

(1) The attenuation value of the optical fiber after the fusion splicing is less than 0.02.

(2) The time for optical fiber fusion splicing should be within 20 minutes.

(3) Color matching of the fiber fusion splicing is correct.

(4) Neat winding and good waterproof treatment of the optical fiber terminal box can be achieved.

4. Safety standards

(1) Teachers and students must correctly wear and use protective gears and tools according to regulations.

(2) All kinds of tools, appliances, equipment and protective gears must be checked before the teaching starts.

(3) When long measuring equipment and device are used, it is necessary to prevent collision with other people and objects; it is forbidden to throw or cast the tools when passing them.

(4) When the power supply is used, the patchboard must be used, and the wires and tools used must be well insulated.

[Requirements of the training]

1. Preparation of tools, meters and devices

One cable connector box, one optical cable toolbox, 20-meter optical cable, and one optical cable fusion splicing machine.

2. Knowledge evaluation points

(1) Classification of optical connecting technology.

(2) Tools for fiber fusion splicing.

(3) The basic principle of fiber fusion splicing.

3. Skills assessment points

(1) Stripping the outer sheath of optical cable.

(2) Fiber cleaning.

(3) Fiber end treatment.

(4) Fiber fusion splicing.

(5) Reinforced protection at the connector.

(6) Coiling fibers.

Task 1 DC Indoor Lightning Protection Box

※ Related knowledge

DC power lightning protection box should also be installed for the outdoor macro coverage. The appearance of the box is shown in Figure 3-1-1. DC power lightning protection box provides lightning protection for RRU power supply, and the installation position should be away from the lower part of the feed line window. The overall dimensions are 390mm×300mm×105mm (height×width×depth).

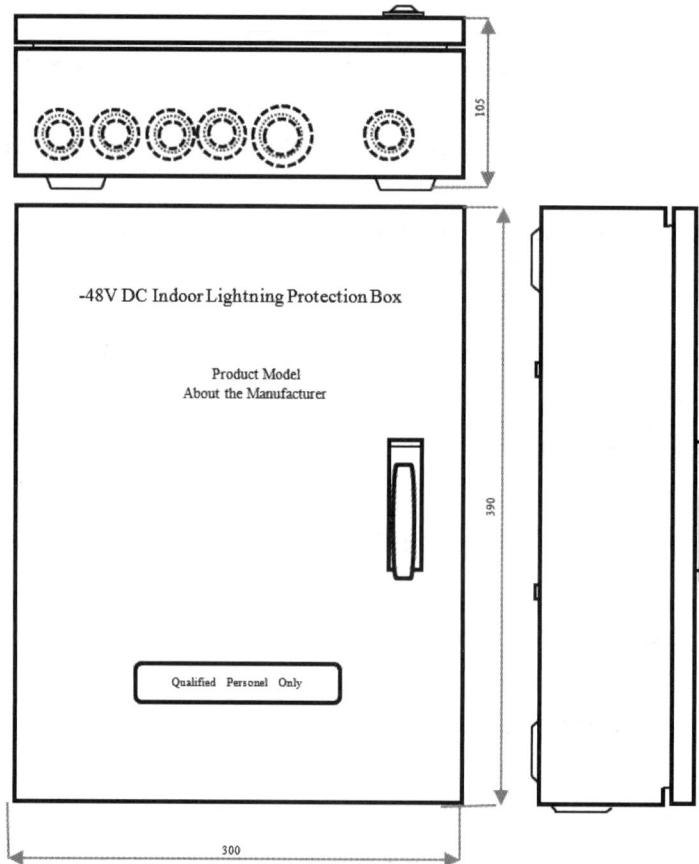

-48V DC Indoor Lightning Protection Box

Product Model
About the Manufacturer

Qualified Personel Only

Figure 3-1-1   Appearance of DC Power Lightning Protection Box

※ Practice

1. Preparation of the tools

Tools required for installation are shown in Table 2-1-1.

2. What to do in this task

The task in this part focuses on the installation of DC indoor power lightning protection box.

3. Training process

1) Fix DC power lightning protection box

The DC power lightning protection box is for wall mounting and is fixed by four M6×60 anchor bolts.

First, confirm the drilling positions of the anchor bolts, then drill four ϕ 8×55 holes perpendicular to the wall, and drill the anchor bolts into the drilled holes, leaving 10 mm exposed for installation. Hang the lightning protection box on four anchor bolts, and then lock them with matched nuts. The installation method is shown in Figure 3-1-2.

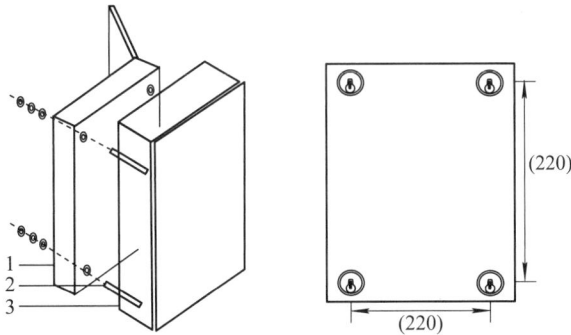

(220)

(220)

1– DC Power Lightning Protection Box; 2– M6×60 Anchor Bolt Subassembly; 3– Wall

Figure 3-1-2　Diagram of the Installation Position and Dimensions

2) Connection of the grounding wire and power line of DC power lightning protection box

The external cables of the DC power lightning protection box should be arranged properly on the horizontal position at the lower part of the lightning protection box. The wiring arrangement of at the bottom of the lightning protection box is as follows: from left to right, grounding wire (16 mm^2), three ways of outdoor RRU power supply cable (double-cored, 6mm^2/core), input power cable (double-cored, 16mm^2/core), remote signaling signal wire (8-cored unshielded wire, 0.2mm^2/core). The terminal of remote signaling backbone node on the lightning protection module is located at the bottom right corner of the box, and the arrangement is shown in Figure 3-1-3.

Figure 3-1-3　Diagram of Interior Modules and Connection

(1) The grounding wire connection of DC power lightning protection box. RVVZ single-cored (16mm^2) (yellow green) wire is also used as the grounding wire of DC power lightning protection box, which can be cut according to the actual need on site. One end should be connected to the connection hole of the grounding terminal of DC power lightning protection box. The outer sheath of the grounding wire should be stripped by 15mm, and then the wire should be plugged into the grounding hole, through crimping method, on the left side of the grounding bus bar of DC power lightning protection box. Then screws should be tightened. The other end should be connected to the indoor or outdoor grounding bar, and it is recommended that it be connected to the outdoor grounding bar by using 25mm^2-M8 copper lug. The requirements for copper lugs making, cable binding and identification tags are the same with that for the main equipment grounding wire.

(2) The power line connection of DC power lightning protection box. Two kinds of power line are used in DC power lightning protection box, namely 2-cored 6mm^2 power line and single-cored 16mm^2 power line. The 2-cored 6mm^2 power line supplies power to RRU, and the single-cored 16mm^2 power line is the input power line of DC power lightning protection box. The location of each terminal is shown in Figure 3-1-3.

For the single-cored 16mm^2 power line:

① Strip the outer sheath with a wallpaper knife to expose the core wire for about 15mm.

② Use a wallpaper knife to cut a small opening on the rubber sheath of the corresponding port at the bottom of the DC power lightning protection box.

③ Use a screwdriver to open the fixed terminal of the power line.

④ Use a screwdriver to loosen the screws of the corresponding power terminal.

⑤ Pass the cable through the rubber sheath.

⑥ Plug the core wire of the power line into the corresponding power terminal, red or black one into 0V, and blue one into –48V.

⑦ Tighten the screws of the power terminal.

⑧ Fix the two 16mm^2 power lines of 0V and –48V with the same power line fixing terminal.

⑨ Use power label tape and the printed sticker at the point 50mm outside the box. The tape and sticker should face the outside of the lightning protection box.

⑩ Strip the outer sheath of the power line at the other end with a wallpaper knife, exposing the core line for about 15mm.

⑪ Use hydraulic pliers to crimp 16mm^2 M8 cable lug onto the core line.

⑫ After the cable lug is crimped, it is necessary to cut the heat-shrink tubing by 50mm and then heat it by using the heat gun at the place where the cable lug and the line are connected;

⑬ The power line should be connected to the corresponding terminal of the power distribution cabinet provided by the operator.

⑭ Use power label tape and the printed sticker at the point 50mm away from the connector. The tape and sticker should face the outside of the power distribution cabinet.

⑮ The requirement for cable binding is the same as that for main equipment power line. For the two-cored 6mm^2 power line:

① Use a wallpaper knife to strip the outer sheath and shielding layer of the power line by about 20 mm.

② Use a wallpaper knife to strip the sheath of each core by about 10 mm.

③ Use a wallpaper knife to remove the outer sheath about 48mm-58mm from the connector of the two-cored 6mm^2, exposing the shielding layer.

④ Use a wallpaper knife to cut a small opening on the rubber sheath of the corresponding port at the bottom of the DC power lightning protection box.

⑤ Use a screwdriver to open the grounding terminal of the shielding layer.

⑥ Use a screwdriver to loosen the screws of the corresponding RRU power supply terminal.

⑦ Pass the line through the rubber sheath.

⑧ Connect the red core wire to the 0V terminal, the blue core wire to the –48V terminal, and then tighten the screws.

⑨ Fix the shielding layer about 73mm away from the connector to the grounding terminal of the shielding layer.

⑩ The aluminum sign of base station (A-RRU or B-RRU or C-RRU) used at the point 50mm from the lightning protection box should be clear in contents, and face outside of the lightning protection box. Bind two 2.5mm×100mm (white) tapes to the power line, with the binding direction of the tapes being the same, and then use diagonal pliers to cut off the redundant tapes along the buckles.

⑪ The other end of the power line should be connected to the RRU.

⑫ The requirement for indoor cable binding is the same as that for main equipment power line.

4. Attentions

⚠Caution:

Before construction is complete, all air switches should be kept off!

After the wires and cables are connected, check whether the connections are correct and firm. If positive, then connect the power supply, and check whether "SPD status indicator" light and "current status indicator" light are green. When everything is normal, close the lightning protection box and put it into operation.

⚠Caution:

The equipment should be installed by authorized professionals; the power must be disconnected during the installation, and it is strictly prohibited to conduct the installation when the equipment is connected to the power, to avoid accidents. The grounding wire should be 25mm2 multi-stranded insulating copper core, and the power connection line should be two-cored 6mm2 insulating copper core. When the status indicator of the lightning arrester in the lightning protection box is red, it indicates that the protection module is invalid, and the lightning protection module needs to be replaced in time. When wires and cables are to be connected, the grounding wire should come first, followed by other wires; when wires and cables are to be removed, the grounding wire should be the last.

5. Conclusion of the task

_____

_____

_____

_____

_____

Task 2　Waterproof Treatment of the Grounding Kit

※ Related knowledge

If the length of feed line exceeds 70m, lightning protection grounding kit must be installed at the middle of the line, and it must be waterproof.

The arrangement of the grounding wire and the feed line should follow a rule: the grounding wire cannot be laid upward from the waterproof place (the joint of the grounding wire and the feed line shielding layer), to prevent water from flowing backward.

※ Practice

1. Preparation of the tools

Tools required for installation are shown in Table 2-1-1.

2. What to do in this task

Conduct waterproof treatment for the grounding kit.

3. Training process

(1) Two products in the following list are used for the waterproof treatment of the grounding kit:

① 2228# waterproof insulating tape: 50mm×37.5mm.

② Super33+ PVC tape: 25mm×10m.

(2) Steps to make grounding kit waterproof:

① Before the work starts, grounding clips, connectors and other areas that need to be covered should be cleaned, to avoid residuals.

② Cut the 2228# waterproof insulating tape by about 30mm and wrap it on the upper end of the grounding wire in advance. The front end of the tape should be closely connected with the end of the cable lug, so as to fill the gap between the grounding wire and the feed line and to prevent water, fog, etc. from entering the connector.

③ Use the Super33+ PVC tape with a width of 25mm to wrap a whole circle in half-overlapping way to cover the whole connector, and the length of wrapping the connector is about 120mm.

⚠Caution:

PVC tape must be used to wrap the object in half-overlapping way. During the whole process of wrapping, it should be tightened and pressed firm. The end should be cut with a knife. After the process is complete, the tape should be pressed by hand to make it completely attached.

④ Use 2228# waterproof insulating tape to cover the whole connector in half-overlapping way, for at least one circle.

⚠Caution:

The 2228# tape must be used to wrap the object in half-overlapping way, and be evenly stretched to about 3/4 of the original width. During the whole process of wrapping, it should be tightened and pressed firm. The end should be cut with a knife. After the process is complete, the tape should be pressed by hand to make it completely attached.

⑤ Use the Super33+ PVC tape with a width of 25mm in half-overlapping way to cover the whole area wrapped with 2228# tape for at least two circles, with the tape extending 3cm-5cm at both ends.

⚠Caution:

PVC tape must be used to wrap the object in half-overlapping way. During the whole process of wrapping, it should be tightened and pressed firm. The end should be cut with a knife. After the process is complete, the tape should be pressed by hand to make it completely attached.

⑥ After wrapping, check the overall condition of the connector to make sure that the tape and appearance are not damaged.

4. Attentions

When using needle-nose pliers, scissors and other tools, students must pay attention to their personal safety, and they must wear protective gloves during the operation, so as to prevent safety accidents that will harm them.

5. Conclusion of the task

_____

_____

_____

_____

_____

Task 3　Introduction to the Classification of

Optical Fiber Fusion Splicing

※ Related knowledge

Fiber fusion splicing machines can be classified as follows.

(1) According to the number of fibers that can be fusion spliced at a time, optical fiber fusion splicing machines can be divided into:

① Single-core fusion splicing machine: which is the most widely used one at present.

② Multi-core fusion splicing machine: which can fusion splice one ribbon optical fiber at one time. It is used mainly for the high-density optical cable in the user transmission line, which can complete the task with heavy load and at a high speed.

(2) According to the types of optical fiber, optical fiber fusion splicing machines can be divided into:

① Multi-mode fusion splicing machine: which cannot be used for the fusion splicing of the single-mode fiber.

② Single-mode fusion splicing machine: which can be used for the fusion splicing of the multi-mode fiber, but the speed is slow.

③ Multi-mode/single-mode fusion splicing machine: which can be used for the fusion splicing of the single or multi-mode fiber through the conversion control machine.

According to the operation mode, optical fiber fusion splicing machines can be divided into, manual(or semi-automatic) fusion splicing machine and automatic fusion splicing machine.

※ Practice

1. Preparation before the task

Prepare fusion splicing equipment and consumables in advance. If the fusion splicing equipment is not available, pictures, videos and other materials are acceptable.

2. What to do in this task

Be acquainted with the fiber fusion splicing and fiber fusion splicing equipment.

1) Application of the optical fiber fusion splicing machine

The optical fiber fusion splicing machine is used mainly for the construction and maintenance of optical cables in optical communication. The general working principle is to melt the cross sections of two optical fibers with high-pressure electric arc, and at the same time smoothly combine the two optical fibers into one with high-precision movable substrate, so as to realize the coupling of optical fiber mode field.

2) The structure and the typical operating keys for optical fiber fusion splicing machine

The structure of typical optical fiber fusion splicing equipment is shown in Figure 3-3-1.

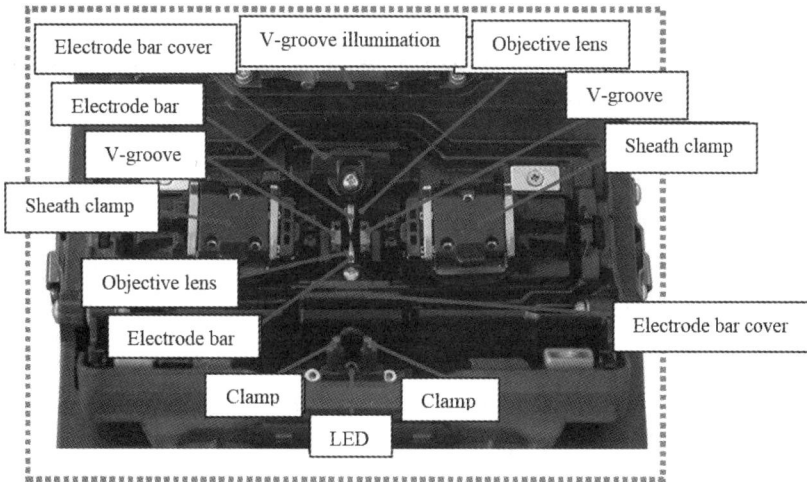

Figure 3-3-1　Structure of Optical Fiber Fusion Splicing Equipment

The typical operating keys for optical fiber fusion splicing equipment are shown in Figure 3-3-2.

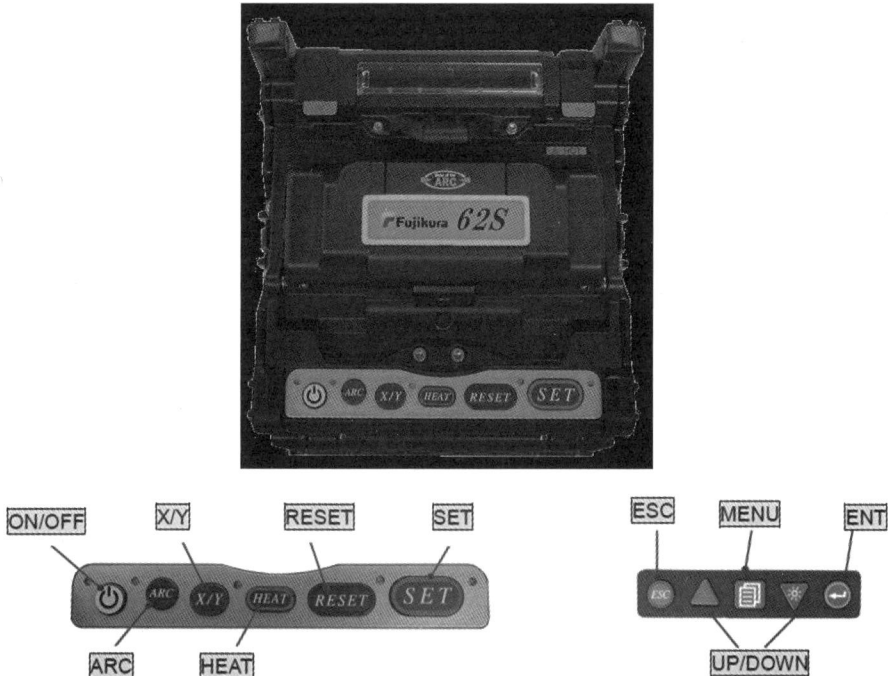

Figure 3-3-2　Typical Operating Keys for Optical Fiber Fusion Splicing Equipment

3) Main steps of optical fiber fusion splicing

The usage of commonly used single-cored fiber fusion splicing machines are basically the same. Specific steps are as follows:

(1) Strip the optical cable, and fix the optical cable to the cable rack. The common optical cables are layer stranding type, skeleton type and central tube type. Different optical cables should be stripped in different ways. After being stripped, the optical cables should be fixed to the cable tray.

(2) The stripped optical fibers should pass through the heat-shrink tubing separately. Fibers with different bundles and colors should be separated and pass through the heat-shrink tubing separately.

(3) Turn on the power supply of the fusion splicing machine and select the appropriate fusion splicing mode. Common specifications of optical fiber contain: SM dispersion-unshifted single mode fiber (ITU-T G.652), MM multi-mode fiber (ITU-T G.651), DS dispersion-shifted single mode (ITU-T G.653), NZ non-zero dispersion-shifted fiber (ITU-T G.655), BI bend-immune fiber (ITU-T G.657) etc. Appropriate fusion modes should be chosen according to different types of optical fiber, while the latest optical fiber fusion splicing machines are equipped with a function that can automatically identify various types of optical fiber.

(4) Prepare the fiber end face. The quality of fiber end face will directly affect the quality of fusion splicing, so the qualified end face must be prepared before the fusion splicing. Strip the coating layer with a special stripping tool, and then rub the bare fiber with alcohol-stained clean linen or cotton for several times. Use a precision optical fiber cleaver to cut the optical fiber. For 0.25mm (the outer coating layer) optical fiber, the cutting length should be 8mm–16mm. For 0.9mm (the outer coating layer) optical fiber, the cutting length can only be 16mm.

(5) Place the fibers. Place the optical fibers in the V-groove of the fusion splicing machine, and carefully place the optical fibers pressing plate and the optical fiber clamps. Set the position of the optical fiber in the pressing plate according to the cutting length of the optical fiber, and correctly put it into the wind screen.

(6) Connect the optical fibers. After pressing the connection key, the fibers gradually move to each other. During the process, a short discharge is generated to clean the fiber surface. When the gap between the fiber end faces is appropriate, the fusion splicing machine stops moving the fibers to each other, sets the initial gap, measures, and displays the cleaving angle. After the initial gap setting is completed, the machine starts the core or clad alignment, and then the machine reduces the gap (the final gap setting). The arc generated by high-voltage discharge melts the left optical fiber into the right optical fiber, and finally the microprocessor calculates the loss and displays the value on the display. If the estimated loss is higher than expected, the machine can discharge again by pressing the discharge key. After the discharge, the machine will still calculate the loss.

(7) Take out the optical fiber and reinforce the fusion splicing joint with a heater. Open the

wind screen, take out the optical fiber from the machine, move the heat-shrink tubing to the position of the joint, put it into the heater for heating, and then take out the optical fiber from the heater after heating. During the operation, do not touch the ceramic part of the heat-shrink tubing and the heater due to the high temperature.

(8) Coil and fix the fiber. Coil the connected optical fiber to the collecting tray. Fix the optical fiber, collecting tray, connector box, and terminal box, etc. Then the operation is complete.

3. Conclusion of the task

_____

_____

_____

_____

_____

Task 4   Optical Fiber Fusion Splicing Practice

※ Related knowledge

The method and basic principle of optical fusion splicing: the high temperature (up to 2,000℃) generated by high voltage (about 3kV) point discharge is used to melt optical fibers, and bakes them together at the same time so that they can be fused into one fiber. Obviously, this type of connection enjoys the best stability.

※ Practice

1. Preparation before the task

Prepare the following tools required for optical fusion splicing, according to Table 3-4-1.

Table 3-4-1   Tools for Fiber Fusion Splicing

Tools	Types (Dimensions)	
Heat-shrink protective tubings	Standard tubing 60mm in length [FP-03] 40mm in length [FP-03 (L = 40)]	
	Miniature tubing 15mm in length [FPS01-400-15]	Miniature tubing 20mm in length [FPS01-900-20]

Continued

Tools	Types (Dimensions)	
Fiber strippers	Primary coating fiber stripper [PS-02]	Nylon coating fiber stripper [JS-01]
Fiber clamps (optional)	Fiber clamp [FH-60-250]	Fiber clamp [FH-60-900]
Fiber cleaver	Fiber cleaver [CT-30]	
Fiber cleaners	Alcohol pot [AP-01] Alcohol (purity > 99%) Dust free paper or gauze	

2. What to do in this task

Optical fiber fusion splicing practice.

3. Training process

1) Preparation of the optical fiber

(1) Strip the fiber coating and clean the bare fiber. Use tools to strip the coating layer by 30mm-40mm from the head, clean the optical fiber with gauze or dust-free paper dipped with anhydrous alcohol. Replace the gauze or paper frequently to ensure the cleaning quality, as shown in Figure 3-4-1.

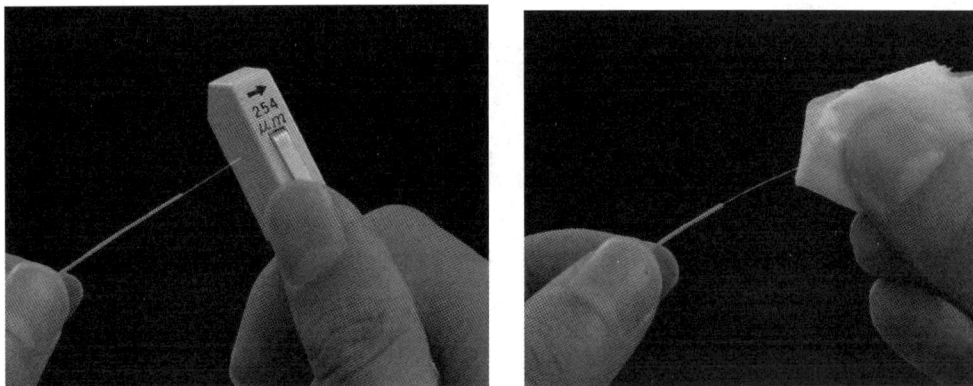

Figure 3-4-1 Strip the Fiber Coating and Clean the Bare Fiber

(2) Use the fiber clamps. Gently press the cleaver arm and slide the latch to unlock the cleaver arm. Push and slide the cleaver substrate to the locked position, and then place the prepared optical fiber on the cleaver. Press the cleaver arm to its bottom, and release it after the sliding cleaver substrate bounces back to its original position. The built-in spring will automatically return it to the open state, as shown in Figure 3-4-2.

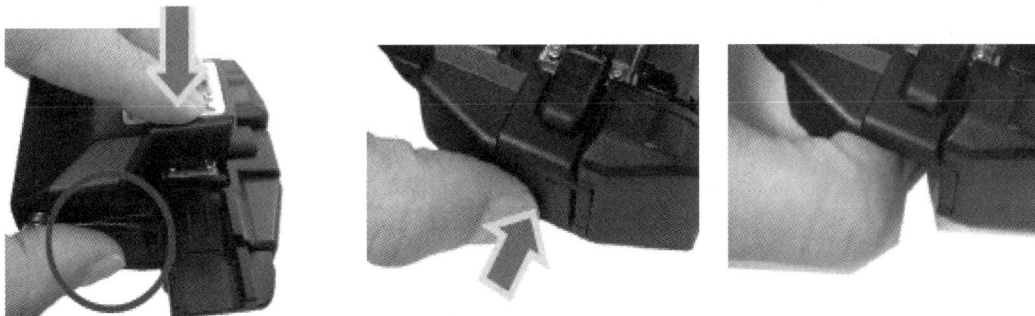

Figure 3-4-2 How to Use Fiber Clamps

2) Put the optical fibers into the fusion splicing machine

(1) Open the wind screen and the clamps.

(2) Put the prepared optical fibers into the V-groove and keep the top of the optical fibers in the area between the electrode bar and the V-groove.

(3) Press the fibers with fingers and close clamps so that fibers will not loosen, making sure the fibers are kept at the bottom of the V-groove. If the fibers are not kept at the correct positions, please replace them as shown in Figure 3-4-3. Make sure that the bottom of the optical fibers should not touch the V-groove and the electrode bar, and the top of the fibers should reach within the frame.

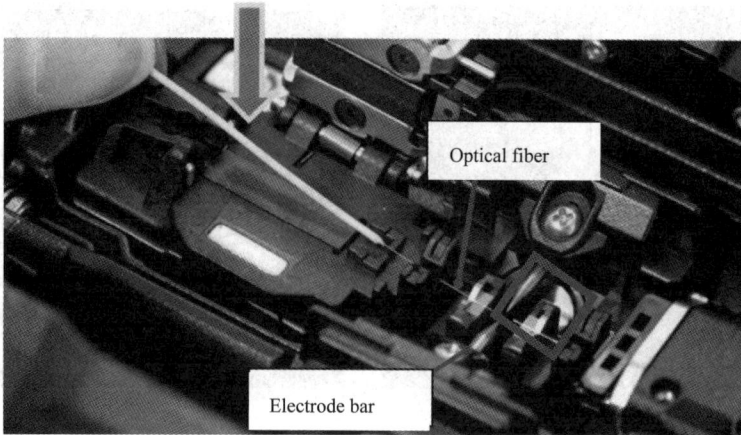

Figure 3-4-3　Correct Position of Optical Fibers in Fusion Splicing Machine

3) Steps of the fusion splicing

For the fusion splicing quality, optical fibers can be monitored by image processing system. However, in some cases, the image processing system cannot detect the failed fusion. Visual inspection through the display is usually a necessary supplementary means. The following instructions describe the standard steps.

(1) The optical fibers are moving closer to each other in the machine. After the cleaning electric arc is released, the optical fibers are then pushed to a proper position and stop. At this time, the display screen of the machine should be what is shown in Figure 3-4-4.

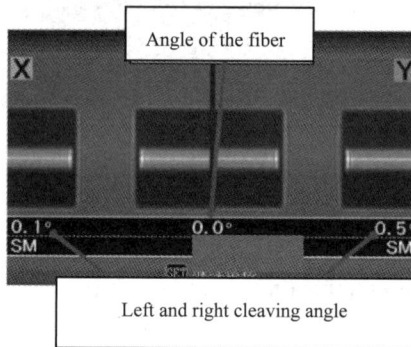

Figure 3-4-4　Image of Optical Fiber under Microscope

(2) Check the cleaving angle and the end face. If the measured value of cleaving angle is greater than the threshold value or fiber debris is detected, the buzzer will ring and an error message will pop up to warn the operator. After checking the optical fibers, make sure that the core should be aligned with the core, or the coating layer should be aligned with the coating layer, and then the fusion process is conducted with electric arc discharge.

(3) After the fusion is completed, the estimated fusion loss will be displayed, as shown in Figure 3-4-5, with influencing factors about the fusion loss listed on the last page. These factors

will be taken into account when the fusion loss is calculated and estimated. The calculation of the fusion loss is based on some spatial parameters, such as the Mode Field Diameter (MFD).

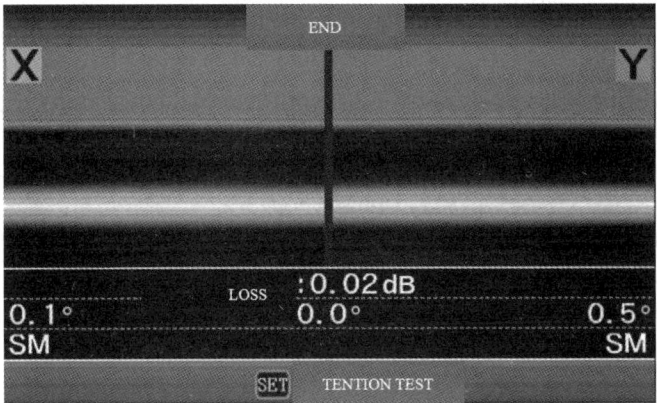

Figure 3-4-5   Estimated Fusion Loss Shown on the Screen

If either value, the detected cleaving angle or the estimated fusion loss, exceeds the preset threshold value, the machine will display an error. If the fiber after being fused is found to be abnormal, such as "too thick" "too thin" or "bubble", the machine will display an error. If there is no error, but a poor fusion is found through visual inspection, it is strongly recommended that the fusing splicing be carried out all over again.

4) Take out the fused optical fiber

(1) Hold the left side of the optical fiber with left hand at the edge of the machine, as shown in Figure 3-4-6.

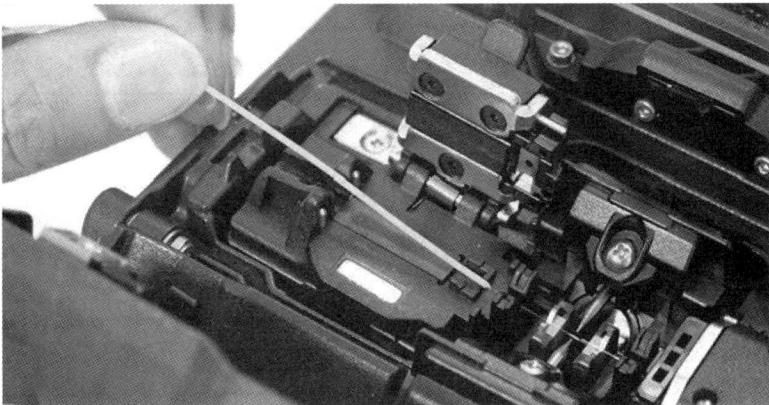

Figure 3-4-6   Take Out the Fiber from the Machine

(2) Open the clamp cover on the right side and slide the heat-shrink tubing from the right side to the fusion point.

(3) Hold the right side of the optical fiber with right hand, and take out the optical fiber.

(4) Hold the fiber with your left hand and move heat-shrink tubing to the fusion point. Make sure that there is no slack in the fusion joint. In addition, avoid the fiber being subjected to excessive pressure, which may cause fracture.

(5) Move the optical fiber down to the heater, as shown in Figure 3-4-7.

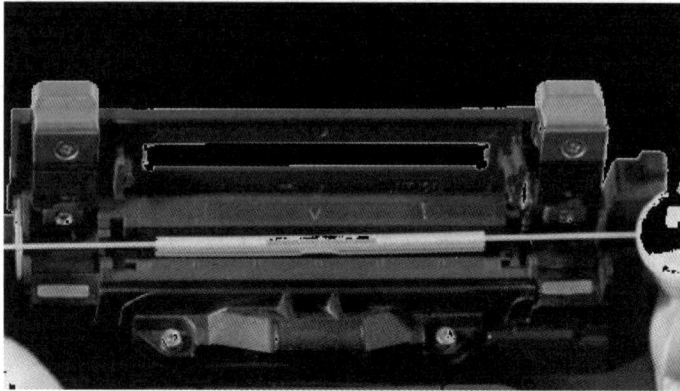

Figure 3-4-7　Put the Optical Fiber Joint with the Heat-shrink Tubing into the Heater

(6) After the optical fiber is put in, the heater cover will automatically close, and the heating cycle will automatically start after the heater cover closes, as shown in Figure 3-4-8.

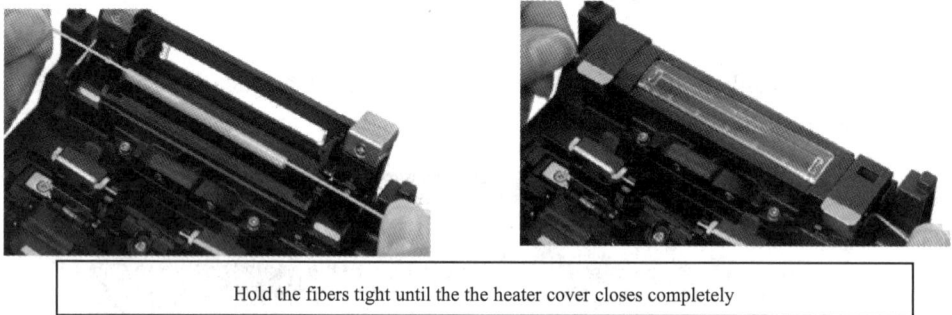

Hold the fibers tight until the the heater cover closes completely

Figure 3-4-8　Heater Begins to Heat the Tubing

(7) After the heating process is completed, the buzzer rings and the heating LED light goes out.

(8) Open the heater cover with appropriate strength. When taking out the heat-shrink tubing from the heater, check whether there are bubbles, dirt and dust inside the heat-shrink tubing.

4. Attentions

First, it is necessary for students to deal with the electricity carefully to prevent electric shock accident when operating the equipment; second, do not touch the heater pipe directly by hand to avoid being scalded; third, do not touch the optical fibers directly to avoid being stabbed.

5. Conclusion of the task

_____

_____

_____

_____

_____

_____